*Solutions Manual to Accompany*

# Revolutions of Geometry

*Solutions Manual to Accompany*

# Revolutions of Geometry

**Michael O'Leary**
College of DuPage
Glen Ellyn, Illinois

WILEY

A JOHN WILEY & SONS, INC., PUBLICATION

Published by John Wiley & Sons, Inc., Hoboken, New Jersey.
Published simultaneously in Canada.



***Library of Congress Cataloging-in-Publication Data:***

O'Leary, Michael
Solutions manual to accompany Revolutions of Geometry.

ISBN 978-0-470-16756-4 (paper)

FSC www.fsc.org MIX Paper from responsible sources FSC® C013604

# CONTENTS

# PREFACE

This manual gives solutions, partial solutions, or hints to 547 exercises from the book *Revolutions of Geometry*. Most odd-numbered problems and a few even-numbered problems have their solutions here. Problems with multiple parts have solutions only to selected parts in this manual. All theorem and figure references refer to the text. References to equations in the solution manual are labeled with symbols such as $(*)$.

A warning for the student: Although helpful, solution manuals can pose temptations that are hard to resist. When working on a problem that has an answer in this manual, the easy path is to look immediately at the answer given before careful thought has been made concerning possible solutions. This avoidance of mental struggle is not conducive to learning. Seek help in these pages only when you are truly stuck or want to confirm a solution.

Michael O'Leary
*College of DuPage*
*June 2010*

*www.revolutionsofgeometry.com*

# CHAPTER 1

# THE FIRST GEOMETERS

## 1.1 EGYPT

**1(a)** Let $a = c$ and $b = d$. Then,

$$\frac{1}{4}(a + c)(b + d) = \frac{1}{4}(2a)(2b) = ab.$$

**1(c)** Recall that the percent error is calculated as

$$\text{percent error} = \left| \frac{\text{true value} - \text{calculated value}}{\text{calculated value}} \right| \times 100.$$

In trapezoid $ABCD$ where $\measuredangle A$ and $\measuredangle B$ are right, $AB = AD = 1$, $BC = 2.12$, and $CD = 1.50$, the trapezoid's actual area is 1.56, and the Egyptian approximation is 1.95. This gives a percent error of 25%.

**3** The correct formula for the truncated cone is

$$V = \frac{\pi h}{12}(D^2 + Dd + d^2).$$

It is derived in a similar manner as the equation for the volume of the truncated pyramid.

**5(a)** .72 foot per every vertical foot.

**5(c)** seked $= b/2h$

**7(a)** Since the apothem of the base is $2\sqrt{3}$, the slant height of the pyramid is $4\sqrt{7}$. Hence, each side of the pyramid is an isosceles triangle of area $8\sqrt{7}$. Because there are six sides to the pyramid, the lateral surface area is $48\sqrt{7}$.

## 1.2 BABYLON

**1(b)** Let $a = 2n + 1$. Then,

$$\frac{a^2+1}{2} = 2n^2 + 2n + 1$$

and

$$\frac{a^2-1}{2} = 2n^2 + 2n.$$

Hence, we find that

$$(2n+1)^2 + (2n^2+2n)^2 = (2n^2+2n+1)^2.$$

**3** $60 = 1,\ 0$
$56861 = 15,\ 47,\ 41$
$4968011 = 23,\ 0,\ 0,\ 11$
$315839601 = 24,\ 22,\ 13,\ 13,\ 21$

**5(a)** $0;\ 21,\ 54$

**5(c)** $3:8$ is $0;\ 22,\ 30$ because the ratio is 3 times $1:8$.

**7** The area is 400 because the square has sides of length 20.

**9** The ratio is

$$\frac{5s_5}{C} = \frac{5 \sin \pi/5}{\pi} \approx .935.$$

## 1.3 CHINA

**1** We will check the formulas for the area of a circle, the volume of the tetrahedron, and the volume of a right circular cylinder.

- The formula for the area of the circle is correct:

$$A = \frac{C}{2} \cdot \frac{d}{2} = \frac{2\pi r}{2} \cdot \frac{2r}{2} = \pi r^2.$$

- Noting that a tetrahedron is a pyramid, we see that the formula is correct because

$$V = \frac{1}{3} \cdot \frac{ab}{2} \cdot h = \frac{abh}{6}.$$

- Since the volume of a right circular cylinder is given by $V = \pi r^2 h$, letting $\pi = 3$ we see that again the formula is correct:

$$V = \frac{C^2 h}{12} = \frac{(2 \cdot 3r)^2 h}{12} = 3r^2 h.$$

**2(b)** 29 feet (treat the length of the vine as the hypotenuse of a right triangle)

**2(d)** 49.5 feet

**2(f)** One method is to draw the triangle so that the legs lie on the $x$- and $y$-axes with the right angle at the origin. Any point on the line $y = x$ is equidistant to the legs, so the center of the circle must lie on this line. Let the hypotenuse be the segment joining $(15, 0)$ and $(0, 8)$. The angle that the hypotenuse makes with the $x$-axis is about 28.072°, which means that the line that bisects this angle is given by $x + 4y = 15$. The intersection of this line with $y = x$ is $(3, 3)$. Hence, the circle that is inscribed within this right triangle has radius 3.

**3** Here is the solution for $n = 5$:

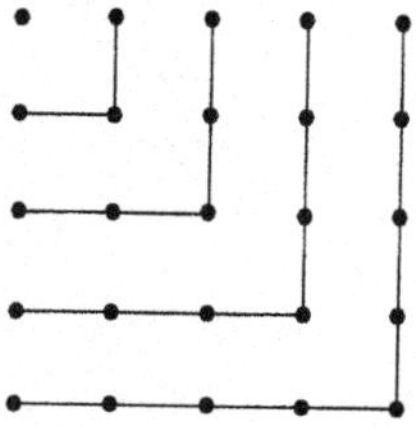

# CHAPTER 2

# THALES

## 2.1 THE AXIOMATIC SYSTEM

**1(a)** Not a proposition. This is a greeting.

**1(c)** A proposition.

**1(e)** Not a proposition. This is a command.

**1(g)** Not a proposition. The word *he* is acting like a variable. In order to consider this sentence to be a proposition, to whom *he* refers must be made clear.

**1(i)** This is equivalent to the sentence *this statement is false*, which is not a proposition.

**2(a)** Deductive. Valid.

**2(c)** Inductive. Not cogent because the conclusion was based on only ten dates.

**2(e)** Deductive. Valid.

**2(g)** Inductive. Cogent.

**2(i)** Deductive. Whether or not this is valid depends on one's understanding of the first sentence. In particular, is the first sentence true if there are no geometers on Jupiter? If so, the argument is invalid. If not, it is valid.

**3** See the solutions given for Exercise 2.

**5** For a good conjecture, what it means for a circle to be *bisected* needs to be clarified.

**7(a)** Proof:

$$\begin{aligned}(a+b)\cdot c &= c(a+b) \text{ [Commutative]}\\ &= ca+cb \text{ [Distributive]}\\ &= ac+bc \text{ [Commutative]}\end{aligned}$$

**7(c)** Proof:

$$\begin{aligned}a+(b+c) &= a+(c+b) \text{ [Commutative]}\\ &= (a+c)+b \text{ [Associative]}\\ &= (c+a)+b \text{ [Commutative]}\end{aligned}$$

**7(e)** Since

$$a+(-1)\cdot a = [1+(-1)]a = 0\cdot a$$

and

$$(-1)\cdot a + a = [(-1)+1]a = 0\cdot a,$$

the additive inverse of $a$ (which is denoted by $-a$) is $(-1)\cdot a$.

**7(g)** The additive inverse of $a+b$ is $(-a)+(-b)$ because

$$\begin{aligned}(a+b)+[(-a)+(-b)] &= (b+a)+[(-a)+(-b)] \text{ [Commutative]}\\ &= b+(a+[(-a)+(-b)]) \text{ [Associative]}\\ &= b+[(a+[-a])+(-b)] \text{ [Associative]}\\ &= b+[0+(-b)] \text{ [Inverse]}\\ &= b+(-b) \text{ [Identity]}\\ &= 0 \text{ [Inverse]}\end{aligned}$$

and $[(-a)+(-b)]+(a+b)=0$.

## 2.2 DEDUCTIVE LOGIC

**1(a)** False.

**1(c)** False.

**1(e)** False.

**1(g)** If this is a proposition, it is false.

**1(i)** True.

**1(k)** False.

**2(a)** *or, if*

**3(a)** Antecedent: *The triangle has two congruent sides.*
Consequent: *It is isosceles.*

**3(c)** Antecedent: *The data are widely spread.*
Consequent: *The standard deviation is large.*

**3(e)** Antecedent: *The system has a solution.*
Consequent: *The system of equations is consistent.*

**5(a)** *The integer is divisible by 4 implies that it is even.*

**5(b)** *The integer is divisible by 4 only if it is even.*

**5(d)** *The integer being even is necessary for it to be divisible by 4.*

**7(a)** Invalid, affirming the consequent

**7(c)** Valid, *Modus Ponens*

**7(e)** Valid, definition of the biconditional

**7(g)** Valid, *Modus Tolens*

**8(a)** *if p then q*

**8(e)** *The integer is not divisible by 4.*

## 2.3 PROOF WRITING

**1(a)** Let $a$ divide $b$. This means that $b = ak$ for some integer $k$. Since we then have $bd = a(kd)$, we see that $a$ also divides $bd$.

**1(c)** Suppose that $a$ divides $b$ and $c$ divides $d$. In other words, $b = ak$ and $d = cl$, where $k$ and $l$ are integers. Then,

$$bd = (ak)(cl) = ac(kl).$$

That is, $ac$ divides $bd$.

**2(a)** Let $n = 2k + 1$ and $m = 2l + 1$ (where $k$ and $l$ are integers). Then,

$$n + m = (2k + 1) + (2l + 1) = 2k + 2l + 2 = 2(k + l + 1).$$

Since $k + l + 1$ is an integer, $n + m$ is even.

**3(a)** Assume that $a = 2k$ and $b = 2l$. We calculate:

$$a^4 + b^4 + 32 = 16k^4 + 16l^4 + 32 = 16(k^4 + l^4 + 2).$$

**3(c)** Let $a^4$ be even but suppose that $a = 2k + 1$. We calculate to find that

$$a^4 = (2k + 1)^4 = 2(8k^4 + 16k^3 + 12k^2 + 4k) + 1,$$

which is odd and contradicts the hypothesis.

**5(a)** First suppose that $a$ is even. This means that $a = 2k$ for some integer $k$. Then, $a^2 = 4k^2$, and this is even. Therefore, we have proven that

*if $a$ is even, then $a^2$ is even.*

To prove the converse, assume that $a$ is odd and write $a = 2k + 1$. Thus,

$$a^2 = (2k + 1)^2 = 2(2k^2 + 2k) + 1.$$

Hence, we have shown that

*if $a$ is odd, then $a^2$ is odd,*

but this is the contrapositive of

*if $a^2$ is even, then $a$ is even.*

**5(c)** Follow the strategy in the solution to Exercise 5(a).

**7** We must prove three implications.

- Assume that $a$ is divisible by 3. In other words, $a = 3k$ for some integer $k$. Hence, $3a = 9k$, so $3a$ is divisible by 9.
- Next, let $3a$ be divisible by 9. Write $3a = 9k$ ($k$ an integer). Hence, $a = 3k$ and $a + 3 = 3(k + 1)$. This means that $a + 3$ is divisible by 3.
- Lastly, suppose that $a + 3$ is divisible by 3, so for some integer $k$ we have that $a + 3 = 3k$. Therefore, $a = 3(k - 1)$, and we see that $a$ is divisible by 3.

**8(a)** Let $a = 0$ or $b = 0$. We have two cases to consider.

- If $a = 0$, then $ab = 0 \cdot b = 0$.
- If $b = 0$, then $ab = a \cdot 0 = 0$.

**8(e)** We will check whether $a(a+1)$ is even under two assumptions:

- First, let $a = 2k$. Then,
$$a(a+1) = 2(2k^2 + k),$$
which is even.

- Now let $a = 2k + 1$. Then,
$$a(a+1) = (2k+1)(2k+2) = 2(2k+1)(k+1).$$
This is also even.

**9(a)** We have two cases to check:

- Suppose that $x \geqslant 0$. This means that $|x| = x$ and $|-x| = -(-x) = x$.
- Now assume that $x < 0$. We see that $|x| = -x$ and $|-x| = -x$. The second equation holds because $-x > 0$.

In both cases we see that $|x| = |-x|$.

**9(c)** Let $a > 0$.

- Assume that $|x| < a$. If $x \geqslant 0$, then $x < a$, and if $x < 0$, then $x > -a$. Combining these, we see that $-a < x < a$.

- Now suppose that $-a < x < a$. If $x \geqslant 0$, then $-a < |x| < a$. On the other hand, if $x < 0$, then $-a < |x| < a$ because $a > -x > -a$. In either case, $|x| < a$.

**11(a)** Since $AB = AB$, we know that $\overline{AB} \cong \overline{AB}$.

**11(b)** Assume that $\overline{AB} \cong \overline{CD}$. This means that $AB = CD$. Since equality is symmetric, $CD = AB$. Hence, $\overline{CD} \cong \overline{AB}$.

**13(a)** Assume that $a \equiv b \pmod{n}$ and $c \equiv d \pmod{n}$. This means that $a - b = nk$ and $c - d = nl$ for integers $k$ and $l$. Calculate:

$$(a+c) - (b+d) = nk + nl = n(k - l).$$

Thus, $a + c \equiv b + d \pmod{n}$.

**13(c)** Using the variables from the solution to Exercise 13(a), we calculate:

$$\begin{aligned} ac - bd &= (nk + b)(nl + d) - bd \\ &= n^2kl + nkd + nlb + bd - bd \\ &= n(nkl + kd + lb). \end{aligned}$$

Therefore, $ac \equiv bd \pmod{n}$.

CHAPTER 3

# PLATO AND ARISTOTLE

## 3.1 FORM

The are no exercises for this section.

## 3.2 CATEGORICAL PROPOSITIONS

**1(a)** Subject = *those destitute of geometry*, predicate = *students of Plato*

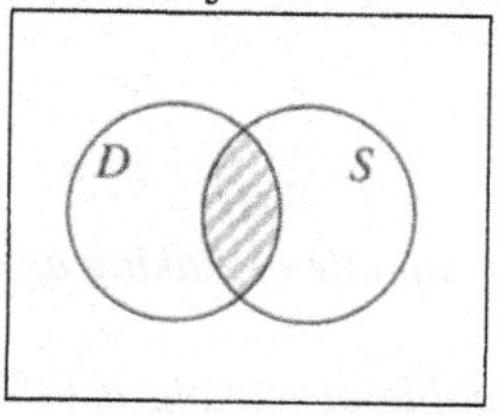


Change the quantity: *Some not destitute of geometry were students of Plato.*
Change the quality: *All destitute of geometry were students of Plato.*

**1(c)** Subject = *geometers*, predicate = *philosophers who excel at mathematics*

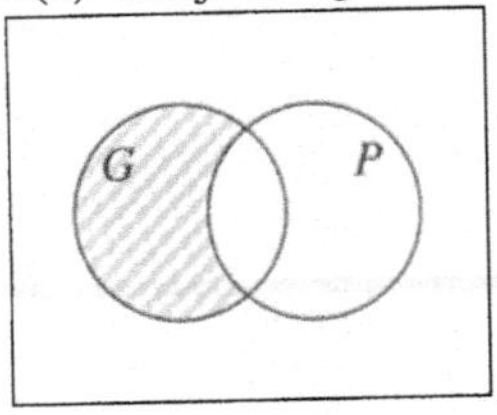


Change the quantity: *Some geometers are philosophers who excel at mathematics.*
Change the quality: *No geometers are philosophers who excel at mathematics..*

**1(e)** Subject = *lines through the point P*, predicate = *lines parallel to l*

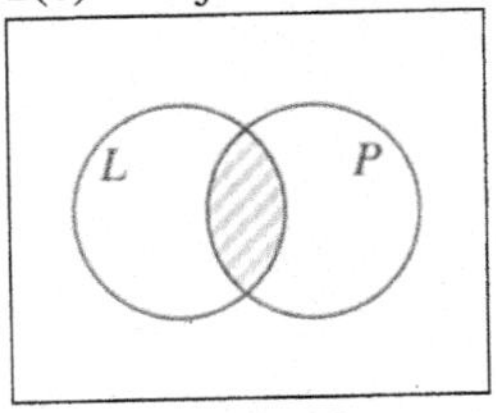


Change the quantity: *Some lines through the point P are parallel to l.*
Change the quality: *All lines through the point P are parallel to l.*

**1(g)** Subject = *people who listened to Thales*, predicate = *Egyptians who preferred to do geometry the old-fashioned way*

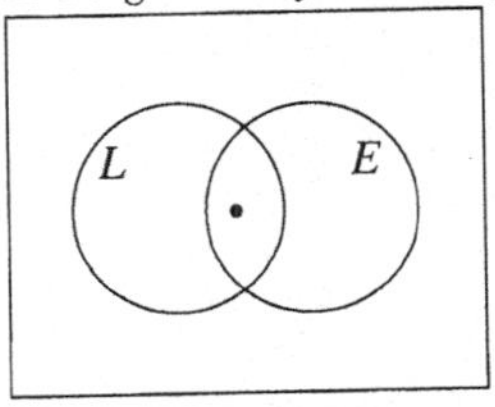


Change the quantity: *All people who listened to Thales were Egyptians who preferred to do geometry the old fashioned way.*
Change the quality: *Some people who did listen to Thales were not Egyptians who preferred to do geometry the old-fashioned way.*

**1(i)** Subject = *Babylonian mathematicians*, predicate = *individuals capable of making astronomical observations*

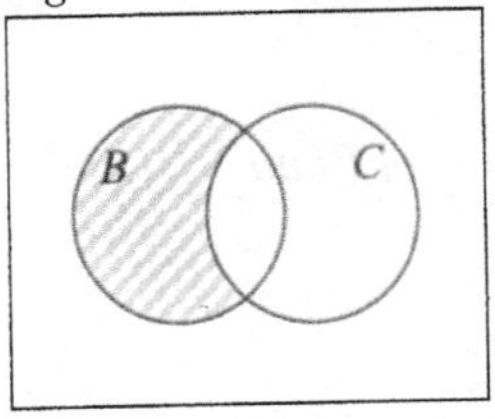


Change the quantity: *Some Babylonian mathematicians were capable of making astronomical observations.*
Change the quality: *No Babylonian mathematician was capable of making astronomical observations.*

**3** See the solutions to Exercise 1.

**5** See the solutions to Exercise 1.

**6(a)** *All B are A*, *No B are A*, not contradictory
The contradictory of *All B are A* is *Some B are not A*, which is the sentence *Some real numbers do not have a square root.*
The contradictory of *No B are A* is *Some B are A*, which is the sentence *Some real numbers have a square root.*

**7(a)** *There exists a real number x such that for all real y*, $y/x = 9$.

**7(c)** *No function has a y-intercept equal to* $-3$.

**7(e)** *Some interval contains a rational number.*

**8(c)** $n = 2$ is a counterexample because 2 is even but $2^2$ is not a multiple of 8.

**9(a)** Valid, subalternation

**9(c)** Invalid

**9(e)** Invalid

**9(g)** Invalid

**9(i)** Valid, converse

## 3.3 CATEGORICAL SYLLOGISMS

**1(a)** Invalid

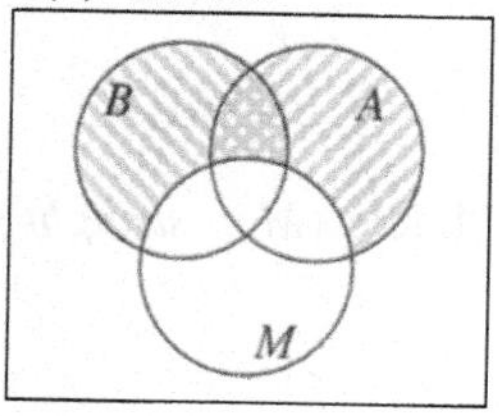


**1(c)** Valid

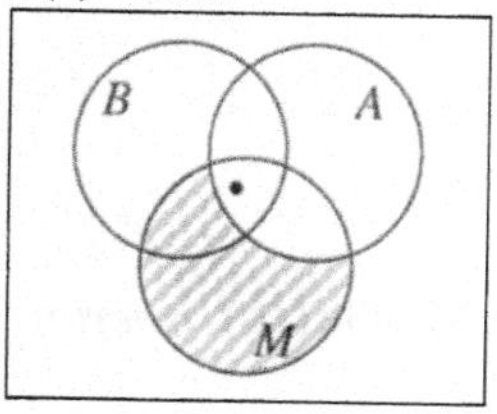

**1(e)** Invalid

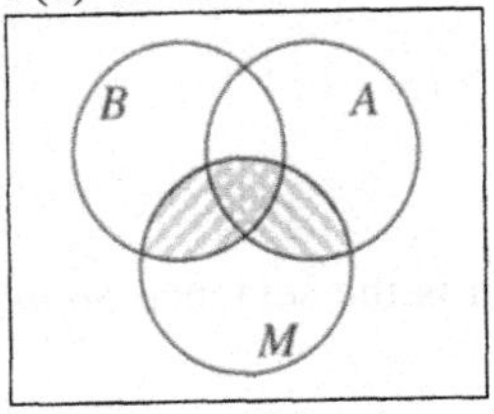


**1(g)** Valid

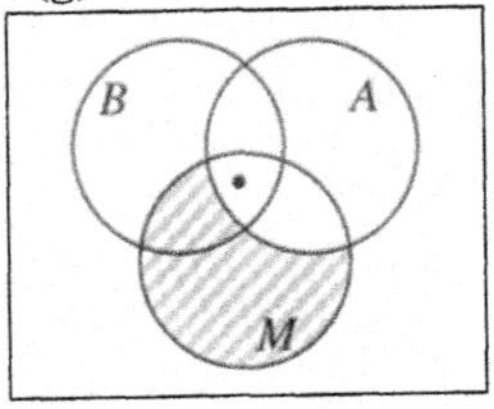


**2(a)** Major ($A$): *mathematicians*, Minor ($B$): *engineers*, Middle ($M$): *geometers*
Invalid

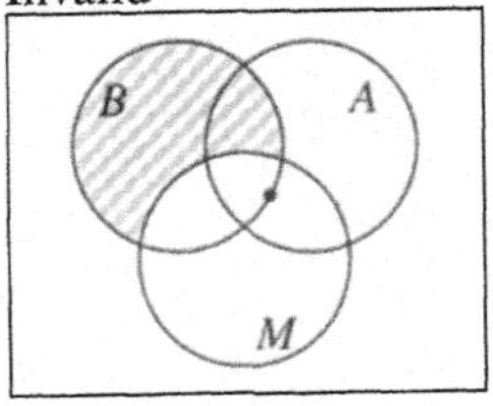


**2(c)** Major ($A$): *circles*, Minor ($B$): *ellipses*, Middle ($M$): *loci with two foci*
Valid

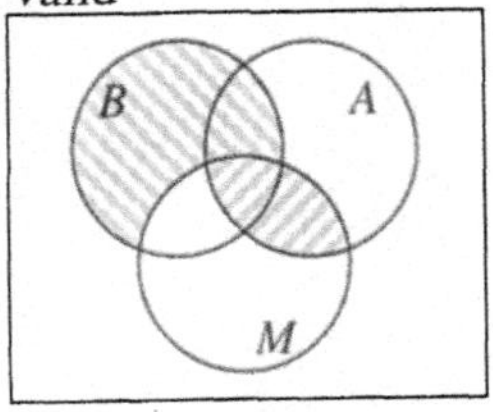


**2(e)** Major ($A$): *parallel lines*, Minor ($B$): *great circles*, Middle ($M$): *straight objects*
Valid

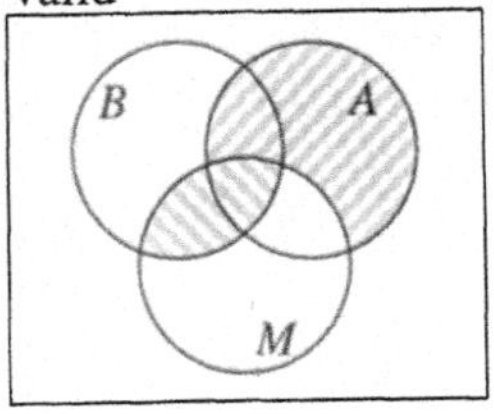


**2(g)** Major ($A$): *interior points of the circle*, Minor ($B$): *points*, Middle ($M$): *exterior points of the circle*

Invalid

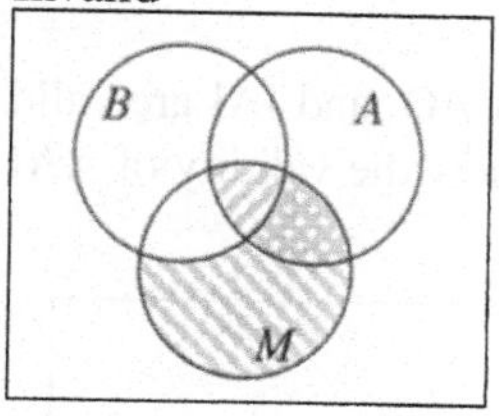


**2(i)** Major ($A$): *graphs of a function*, Minor ($B$): *line segments*, Middle ($M$): *curves*
Invalid

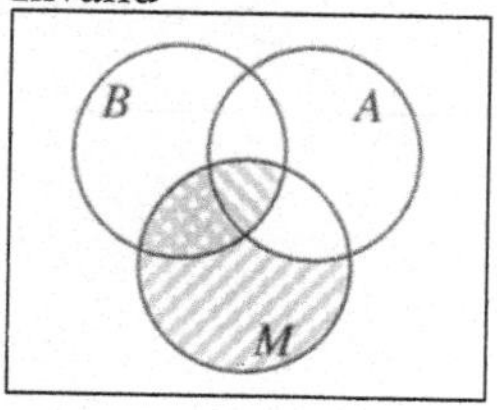


**3** See the solutions to Exercise 2.

## 3.4 FIGURES

**1(a)** *All M are A, No B are M ∴ No B are A*
Figure 1, AEE, invalid

**1(c)** *No M are A, All B are M ∴ No B are A*
Figure 1, EAE, valid

**1(e)** *Some A are M, No M are B ∴ Some B are not A*
Figure 4, IEO, invalid

**1(g)** *Some M are A, All M are B ∴ Some B are A*
Figure 3, IAI, valid

**1(i)** *All A are M, No B are M ∴ No B are A*
Figure 2, AEE, valid

**2** See the solutions to Exercise 1.

**3(a)** *All M are A, All B are M ∴ All B are A*

**3(c)** *No M are A, All M are B ∴ Some B are not A*

**3(e)** *All M are A, Some M are B ∴ Some B are A*

**3(g)** *No A are M, All B are M ∴ No B are A*

**3(i)** *No M are A, All B are M ∴ No B are A*

**3(k)** *No A are M, All B are M* ∴ *Some B are not A*

**5** The following Venn diagrams show that the moods AAI, EAO, and IAI are valid in the third figure. Notice that existential import is used to show the validity of AAI and EAO.

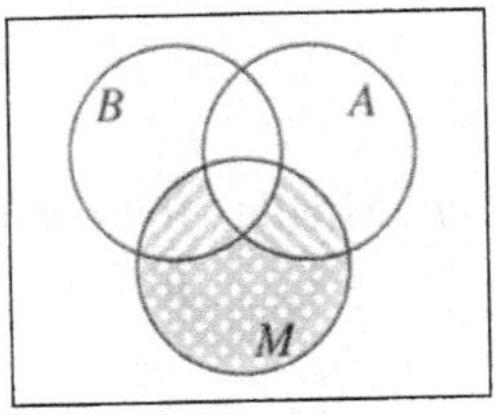


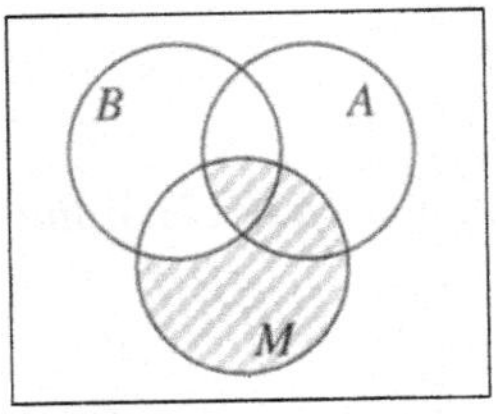


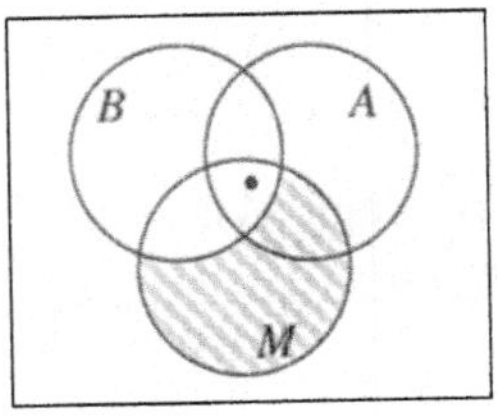

# CHAPTER 4

# PYTHAGORAS

## 4.1 NUMBER THEORY

**1(a)** The equation for the $n$th triangular number is

$$t_n = \frac{n(n+1)}{2}.$$

**1(d)** The equation for the $n$th tetrahedral number is

$$T_n = \frac{n(n+1)(n+2)}{6}.$$

**3** The first four oblong numbers are:

**5(a)** Here is the algebraic proof that $p_n = n^2 + t_{n-1}$:

$$n^2 + t_{n-1} = \frac{2n^2}{2} + \frac{n(n-1)}{2} = \frac{3n^2 - n}{2} = p_n.$$

*Revolutions of Geometry.* By Michael O'Leary

The last equation uses the formula from Exercise 5(b). The following diagram illustrates the formula:

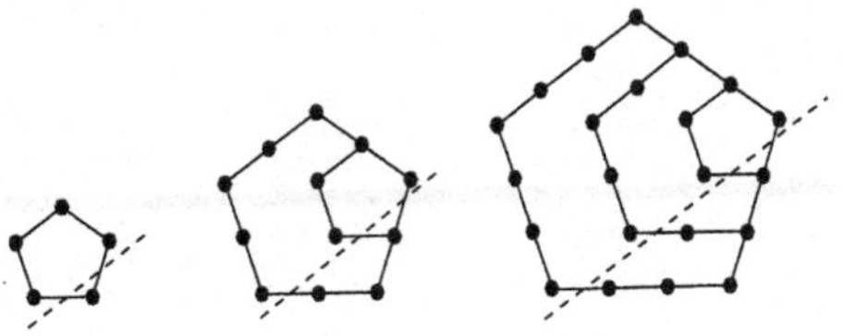

**5(b)** Another formula is derived as follows:

$$p_n = 3(n-1) + 3(n-2) + \cdots + 1 + n = 3\left(\frac{n[n-1]}{2}\right) + n = \frac{3n^2 - n}{2}.$$

**7** Assume that $a > b$ and $c > 0$. Then,

$$\begin{aligned} ac &> bc, \\ ab + ac &> ab + bc, \\ a(b+c) &> b(a+c), \\ \frac{a}{b} &> \frac{a+c}{b+c}. \end{aligned}$$

**9** The two means are the arithmetic and the harmonic. The proportion holds because

$$\left(\frac{a+b}{2}\right)\left(\frac{2ab}{a+b}\right) = ab.$$

**11(a)** Since $x$, $y$, and $z$ are the first three terms of a geometric sequence, there exists $r$ such that $y = rx$ and $z = r^2x$. Therefore,

$$\begin{aligned} a &= x(2 + 3r + r^2), \\ b &= x(2r + r^2), \\ c &= x(r + r^2). \end{aligned}$$

Because

$$a(b-c) = rx^2(2 + 3r + r^2)$$

and

$$c(a-b) = rx^2(2 + 3r + r^2),$$

we conclude that

$$\frac{a-b}{b-c} = \frac{a}{c}.$$

**13** Find $\overline{MN}$ and $\overline{OP}$ so that $AB = kMN$, $CD = lMN$, $A'B' = mOP$, and $C'D' = nOP$ for some positive integers $k$, $l$, $m$, and $n$. Then,

$$AB + A'B' = kMN + mOP \text{ and } CD + C'D' = lMN + nOP.$$

Because of these two equations, there is no reason that $\overline{EF}$ and $\overline{GH}$ must be commensurable.

## 4.2 THE PYTHAGOREAN THEOREM

**1** We assume the following: angles that look right are right, the large square on the left has the same dimensions as the one on the right, and the dimensions of the eight triangles are identical, with $a$ and $b$ being the measures of the legs and $c$ the measure of the hypotenuse. Let $S$ be the area of the large square and $T$ be the area of one of the triangles. Based on the figure on the left,

$$S - 4T = c^2,$$

and based on the figure on the right,

$$S - 4T = a^b + b^2.$$

Thus, $a^2 + b^2 = c^2$.

**3** We will prove the second equation. Let $(x_1, y_1, z_1)$ and $(x_2, y_2, z_2)$ be two points in three dimensions:

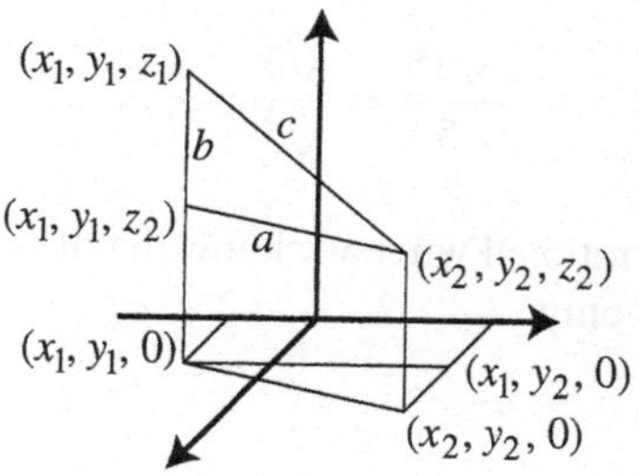


The distance $c$ is what we need to find. Observe that $b = |z_1 - z_2|$ and

$$a = \sqrt{(x_1 - x_2)^2 + (y_1 - y_2)^2}.$$

We used the formula from the first part of this exercise to compute $a$. We use it again to find $c$:

$$\begin{aligned} c &= \sqrt{\left(\sqrt{(x_1 - x_2)^2 + (y_1 - y_2)^2}\right)^2 + |z_1 - z_2|^2} \\ &= \sqrt{(x_1 - x_2)^2 + (y_1 - y_2)^2 + (z_1 - z_2)^2}. \end{aligned}$$

**5** We will follow the method from the section. Suppose that $\sqrt{3} = a/b$ with $a/b$ reduced to lowest terms. Then, $3b^2 = a^2$. This means that 3 divides $a^2$. From this we may conclude that 3 divides $a$. If it did not, then $a$ divided by 3 will have a remainder of 1 or 2. This would mean that $a = 3k + 1$ or $a = 3k + 2$. In both cases $a^2$ would have a remainder of 1 when divided by 3 because

$$(3k + 1)^2 = 9k^2 + 6k + 1 = 3(3k^2 + 2k) + 1$$

and

$$(3k + 2)^2 = 9k^2 + 12k + 4 = 3(3k^2 + 4k + 1) + 1.$$

This is impossible, so we conclude that we can write $a = 3l$. So, $3b^2 = 9l^2$. Hence, 3 divides $b^2$, which means that 3 divides $b$. This contradicts the fact that the original fraction was in lowest terms. Hence, $\sqrt{3}$ is irrational.

**7** The numbers $\sqrt{3}$ and $\sqrt{5}$ are commensurable in square because

$$3 = 45\left(\frac{1}{15}\right)$$

and

$$5 = 75\left(\frac{1}{15}\right).$$

However, if there existed a length $a$ such that $\sqrt{3} = ka$ and $\sqrt{5} = la$ with $k$ and $l$ integers, we may conclude that

$$\frac{\sqrt{15}}{5} = \frac{\sqrt{3}}{\sqrt{5}} = \frac{k}{l}.$$

This means that $\sqrt{15}/5$ is rational when we know that it is not. Thus, $\sqrt{3}$ and $\sqrt{5}$ are commensurable in square only.

## 4.3 ARCHYTAS

**1(a)** Given $\overline{AB}$ such that $AB = a$, construct $\overline{AC}$ perpendicular to $\overline{AB}$ with $AC = a$. The hypotenuse of this triangle has length $a\sqrt{2}$.

**1(b)** The mean proportional between $m = a$ and $n = 2a$ is $a\sqrt{2}$.

**3(a)** Let $a = AB = BC$. Since $\sphericalangle ABC$ is inscribed within a semicircle, $\triangle ABC$ is a right triangle by Thales' Theorem. Hence, $AC = a\sqrt{2}$. Now notice that the area of the large semicircle is $\pi a^2/2$ and the area of each small semicircle is $\pi a^2/4$. Therefore, the sum of the areas of the semicircles on $\overline{AB}$ and $\overline{BC}$ equals the area of the semicircle on $\overline{AC}$.

**5(a)** The bisector of $\sphericalangle ABC$.

**5(c)** The perpendicular bisector of the segment joining the given points.

**5(e)** A circle.

**7** The line $l$ is the locus of points described by $P$:

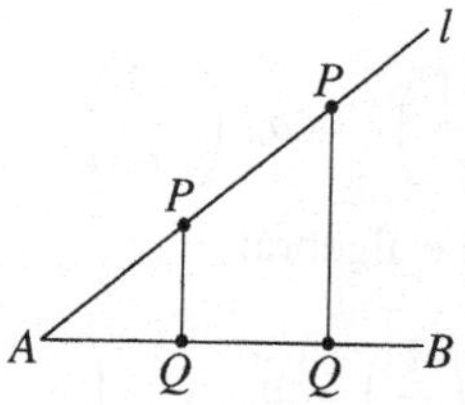


**9** We are given that $AB = a$ and $AC = b$. By Thales' Theorem, $\measuredangle ABC$ is right, so

$$BC = b^2 - a^2.$$

The cone constructed in Archytas's solution has height $AC$ and base with radius $CD$. Since $\triangle ABC \sim \triangle ACD$,

$$\frac{AB}{BC} = \frac{AC}{CD},$$

and we are able to calculate the value of the radius as

$$CD = \frac{b^3 - a^2 b}{a}.$$

Hence, the volume of the cone is

$$V = \frac{\pi}{3a^2}(b^7 - 2a^2 b^5 + a^4 b^3).$$

**11** We will use the Centroid Theorem of Pappus.

- The area of the circle is $A = \pi r^2$. The centroid of the curve is $(a, 0)$. As the circle is rotated about the $y$-axis, the centroid travels a distance of $d = 2\pi a$. Hence,

$$\text{volume(torus)} = Ad = 2\pi^2 a r^2.$$

- The length of the generating curve of the torus is $s = 2\pi r$. Therefore,

$$\text{area(torus)} = sd = 4\pi^2 a r.$$

## 4.4 THE GOLDEN RATIO

**1** First,

$$\frac{1}{\tau} = \frac{2}{\sqrt{5}+1} \cdot \frac{\sqrt{5}-1}{\sqrt{5}-1} = \frac{2\sqrt{5}-2}{4} = \frac{\sqrt{5}-1}{2}.$$

Also, by the quadratic formula, the negative solution to $x^2 - x - 1 = 0$ is

$$x = \frac{1-\sqrt{5}}{2} = -\frac{1}{\tau}.$$

**3** We use

$$\left(\frac{-1}{\tau}\right)^n = a_n\left(\frac{-1}{\tau}\right) + a_{n-1}$$

as our starting point. First, some algebra:

$$a_n\left(\frac{1}{\tau}\right) = a_{n-1} - \left(\frac{-1}{\tau}\right)^n,$$

$$a_n\left(\frac{1}{\tau}\right) = \tau^n - a_n\tau - \left(\frac{-1}{\tau}\right)^n,$$

$$a_n\left(\tau + \frac{1}{\tau}\right) = \tau^n - \left(\frac{-1}{\tau}\right)^n.$$

This confirms the equation because

$$\tau + \frac{1}{\tau} = \frac{\sqrt{5}+1}{2} + \frac{\sqrt{5}-1}{2} = \sqrt{5}.$$

**5** Take the equation

$$x = \sqrt{1+\sqrt{1+\sqrt{1+\sqrt{1+\cdots}}}}$$

and square both sides to find

$$x^2 = 1 + \sqrt{1+\sqrt{1+\sqrt{1+\cdots}}}.$$

Therefore, subtracting the equations gives $x^2 - x = 1$. Since $\tau$ is the only positive solution to this equation, $x = \tau$.

**7** Since $FA \cdot FE = AC^2$ and $\overline{AC} \cong \overline{AE}$,

$$\frac{FE}{AE} = \frac{FA \cdot FE}{FA \cdot AE} = \frac{AC^2}{FA \cdot AE} = \frac{AE^2}{FA \cdot AE} = \frac{AE}{FA}.$$

**9(a)** The length of the curve is given by the expression

$$\int_0^{\theta_0} \sqrt{a^2e^{2b\theta} + a^2b^2e^{2b\theta}}\, d\theta = \int_0^{\theta_0} e^{b\theta}\sqrt{a^2 + a^2b^2}\, d\theta = \frac{(e^{b\theta_0} - 1)\sqrt{a^2 + a^2b^2}}{b},$$

which is finite.

**11** Let $ABCDE$ be a regular pentagon and let $G$ be the midpoint of $\overline{DC}$. Join $\overline{EB}$ and $\overline{AD}$. When $A$ and $G$ are joined, the resulting segment intersects $\overline{EB}$ at its midpoint, which we identify as $F$. We will show that $FG/AF = \tau$ for the case when the sides of the pentagon are unit length. By Exercise 10,

$$EB = AD = \tau.$$

Therefore, by the Pythagorean Theorem,

$$AG = \sqrt{\tau^2 - \frac{1}{4}}$$

and

$$AF = \sqrt{1 - \frac{\tau^2}{4}}.$$

Hence,

$$\frac{AG^2}{AF^2} = \frac{10 + 4\sqrt{5}}{5 + \sqrt{5}} = \frac{3 + \sqrt{5}}{2} = \tau^2.$$

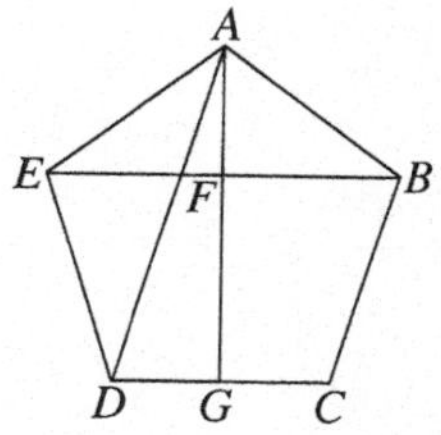


**13** Construct regular pentagon $A_1A_2A_3A_4A_5$ circumscribed within a circle. Bisect each of the five intercepted arcs, resulting in points $B_1$, $B_2$, $B_3$, $B_4$, and $B_5$. Then polygon $A_1B_1A_2B_2A_3B_3A_4B_4A_5B_5$ is a regular decagon.

**15** Since $1 = AD = 2CD + BC$ and $BC = 1/\tau = (\sqrt{5} - 1)/2$,

$$CD = \frac{1 - 1/\tau}{2} = \frac{1}{2} - \frac{1}{2\tau} = \frac{3 - \sqrt{5}}{4},$$

so

$$CD^2 = \left(\frac{3 - \sqrt{5}}{4}\right)^2 = \frac{14 - 6\sqrt{5}}{16} = \frac{7 - 3\sqrt{5}}{8}.$$

Next,

$$\frac{GF}{CD} = \frac{DF}{ED} = \tau,$$

so since $\tau = (\sqrt{5} + 1)/2$,

$$GF^2 = \left(\frac{\sqrt{5} + 1}{2} \cdot \frac{3 - \sqrt{5}}{4}\right)^2 = \left(\frac{\sqrt{5} - 1}{4}\right)^2 = \frac{3 - \sqrt{5}}{8}.$$

Because $1/\tau^2 = (3 - \sqrt{5})/2$,

$$ED^2 = EI^2 - ID^2 = \frac{3 - \sqrt{5}}{2} - \frac{1}{4} = \frac{5 - 2\sqrt{5}}{4}.$$

Next, we know that $EC^2 = ED^2 + CD^2$, so

$$EC^2 = \frac{5 - 2\sqrt{5}}{4} + \frac{7 - 3\sqrt{5}}{8} = \frac{3 - \sqrt{5}}{8}.$$

Finally,

$$DG^2 = \tau^2 EC^2 = \frac{3 + \sqrt{5}}{2} \cdot \frac{3 - \sqrt{5}}{8} = \frac{1}{4},$$

so $DG = 1/2$.

# CHAPTER 5

# EUCLID

## 5.1 THE ELEMENTS

**1(a)** We could argue that there is no exact example of a point in the physical world if what is meant by *point* is a zero-dimensional object. If we allow for approximations, a pixel on a computer monitor could be considered an example of a point.

**5** An example that you could give would be a term that you commonly use only among your family or friends. The meaning of such a term would be understood by all involved, yet no one has given (or possible even could give) a definition for the term.

**7** An example of the first common notion is

$$\text{if } \triangle ABC \cong \triangle GHI \text{ and } \triangle DEF \cong \triangle GHI, \\ \text{then } \triangle ABC \cong \triangle GHI.$$

An example of the third common notion is

$$\text{if } \mathrm{m}\measuredangle ABC = \mathrm{m}\measuredangle A'B'C' \text{ and } \mathrm{m}\measuredangle ABD = \mathrm{m}\measuredangle A'B'D', \\ \text{then } \mathrm{m}\measuredangle ABC - \mathrm{m}\measuredangle ABD = \mathrm{m}\measuredangle A'B'C' - \mathrm{m}\measuredangle A'B'D'.$$

*Revolutions of Geometry.* By Michael O'Leary

provided that $D$ is on the interior of $\sphericalangle ABC$ and $D'$ is on the interior of $\sphericalangle A'B'C'$. Using the same angles, the following is an example of the fifth common notion:

$$m\sphericalangle ABC > m\sphericalangle ABD.$$

## 5.2 CONSTRUCTIONS

**1** We must show that congruence of angles is reflexive, symmetric, and transitive. We will use the fact that equality of numbers is an equivalence relation.

- $\sphericalangle A \cong \sphericalangle A$ because $m\sphericalangle A = m\sphericalangle A$.
- Suppose that $\sphericalangle A \cong \sphericalangle B$. This means that $m\sphericalangle A = m\sphericalangle B$. Therefore, we conclude that $m\sphericalangle B = m\sphericalangle A$, so $m\sphericalangle B \cong m\sphericalangle A$.
- Let $\sphericalangle A \cong \sphericalangle B$ and $\sphericalangle B \cong \sphericalangle C$. This translates as $m\sphericalangle A = m\sphericalangle B$ and $m\sphericalangle B = m\sphericalangle C$. Hence, $m\sphericalangle A = m\sphericalangle C$, and then $\sphericalangle A \cong \sphericalangle C$.

**3** Begin by drawing segments that represent the lengths of $a$, $b$, and $c$.

- $2a$: On a line segment, lay off two copies of $a$ using a compass.

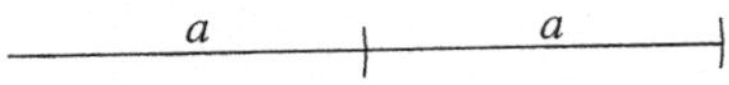


- $a - b$: On a line segment of length $a$, copy a segment of length $b$. The difference is $a - b$.

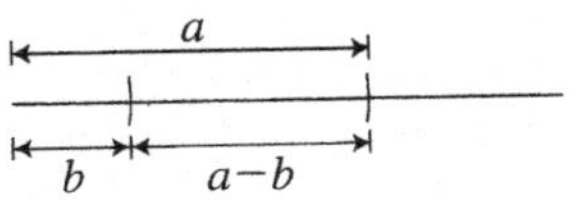


- $c/4$: Copy a segment of length $c$. Find its midpoint and then find the midpoints of the two resulting segments. The resulting four segments each have length $c/4$.

**5** Use the construction of perpendicular lines.

**7(a)** Construct square $ABCD$. Then $\triangle ABD$ is a 45-45-90 triangle.

**7(c)** Construct a right $\sphericalangle ABC$. Pick a point $D$ on the exterior of $\sphericalangle ABC$ such that $m\sphericalangle DBC$ is less than two right angles. Then, $\sphericalangle DBC$ is obtuse. Let $E$ be any point on side $\overline{BD}$. Pick $F$ on $\overline{BC}$ (produced if necessary) such that $\overline{BE} \cong \overline{BF}$. Then, $\triangle EBF$ is an isosceles triangle with an obtuse angle.

**9** Choose a point $O$ on a horizontal line segment to represent 0. Let $A$ also lie on the segment to the right of $O$. At $A$, place the number 1. Then at intervals of length $OA$ on the right and the left place integers in sequence, creating a number line. Construct an isosceles right triangle with legs of length $OA$. Its hypotenuse has length $\sqrt{2}$. Copy the hypotenuse on the number line with one endpoint at $O$ and the other endpoint $B$ to the right of $O$. The point $B$ represents $\sqrt{2}$ on the number line.

## 5.3 TRIANGLES

**1** By hypothesis, $\measuredangle CAB \cong \measuredangle CBA$ and $\measuredangle CDA \cong \measuredangle CDB$. Since $\overline{CD}$ is congruent to itself, $\triangle CDA \cong \triangle CDB$ by AAS. Then, $\measuredangle ACD \cong \measuredangle BCD$ by the definition of congruent triangles.

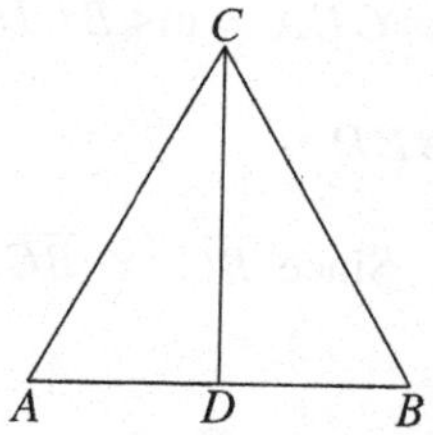


**3** Let $\overline{AC}$ and $\overline{CB}$ intersect $\overline{DE}$ at $C$. Assume that $\measuredangle ACD$ and $\measuredangle DCB$ are supplementary. In order to obtain a contradiction, let $A$, $C$, and $B$ not be collinear. This means that these three points form $\triangle ABC$. However, this triangle has internal angle sum greater than two right angles, contradicting Theorem 5.3.12.

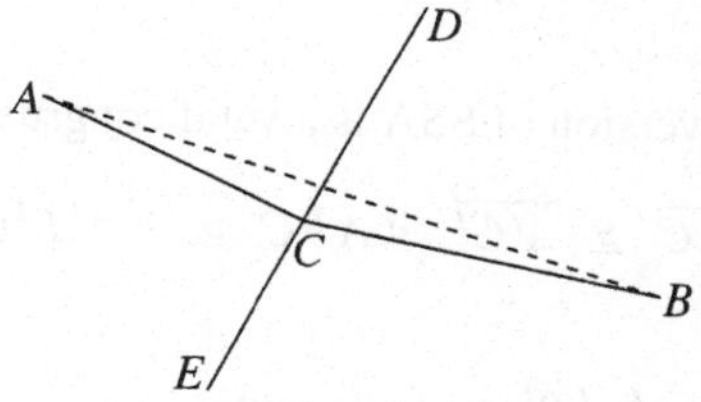


**5(a)** By the Triangle Inequality,

$$BE < BA + AE.$$

Then, $BE + EC < BA + AE + EC$, so we have

$$BE + EC < BA + AC.$$

Similarly, we can prove that

$$BD + DC < BE + EC.$$

Hence,

$$BD + DC < BA + AC.$$

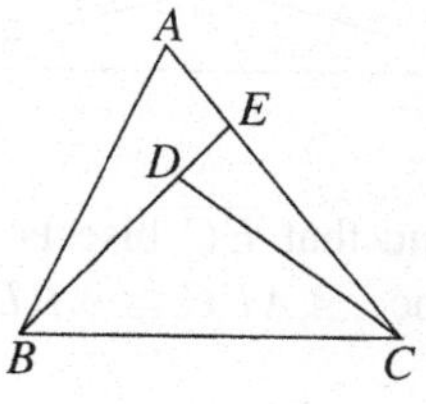

**7** Let $\overline{AB}$ and $\overline{CD}$ meet at $E$. We know that

$$m\sphericalangle CEA + m\sphericalangle AED = m\sphericalangle BED + m\sphericalangle AED$$

since these are two pairs of supplementary angles. Thus,

$$m\sphericalangle CEA = m\sphericalangle BED,$$

which means that $\sphericalangle CEA \cong \sphericalangle BED$.

**9** Let $\sphericalangle B \cong \sphericalangle C$ in $\triangle ABC$. Since $\overline{BC} \cong \overline{BC}$, $\triangle ABC \cong \triangle ACB$ by ASA. Therefore, $\overline{AB} \cong \overline{AC}$.

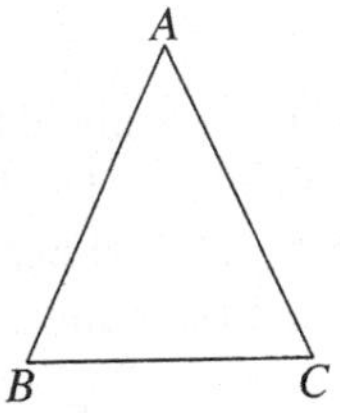


**11** The following strong version of SSA is a valid congruence test:

If $\overline{AB} \cong \overline{A'B'}$, $\overline{AC} \cong \overline{A'C'}$, $\sphericalangle ABC \cong \sphericalangle A'B'C'$, and one of the following hold:

- $\sphericalangle ACB$ and $\sphericalangle A'C'B'$ are both acute,
- $\sphericalangle ACB$ and $\sphericalangle A'C'B'$ are both obtuse, or
- $\sphericalangle ACB$ or $\sphericalangle A'C'B'$ is a right angle,

then $\triangle ABC \cong \triangle A'B'C'$.

**13** By hypothesis, $\overline{AB} \cong \overline{AC}$ and $\overline{AD} \cong \overline{AE}$. Since $\sphericalangle A$ is congruent to itself, $\triangle ABE \cong ACD$ by SAS. Therefore, $\overline{BE} \cong \overline{DC}$. Furthermore, $\overline{BC} \cong \overline{BC}$, and by segment subtraction $\overline{BD} \cong \overline{EC}$. Therefore, $\triangle DBC \cong ECB$ by SSS.

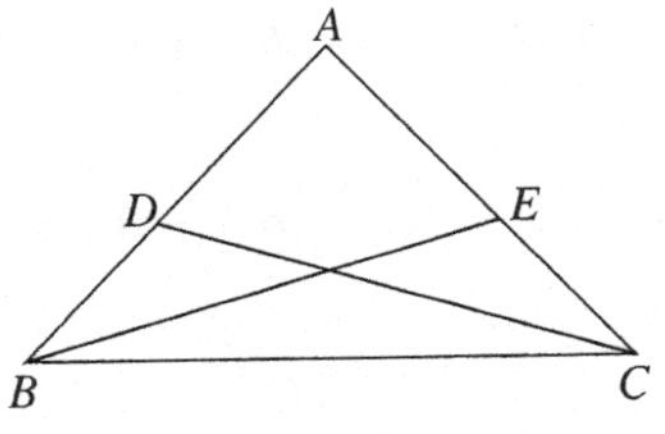


**15** Assume that $\overline{DC} \cong \overline{FC}$ and that $\overline{EC}$ bisects $\sphericalangle ACB$. Because $\overline{EC} \cong \overline{EC}$, $\triangle DCE \cong \triangle FCE$ by SAS. Hence, $\sphericalangle AFC \cong \sphericalangle BDC$, which shows that $\triangle AFC \cong \triangle BDC$ by ASA.

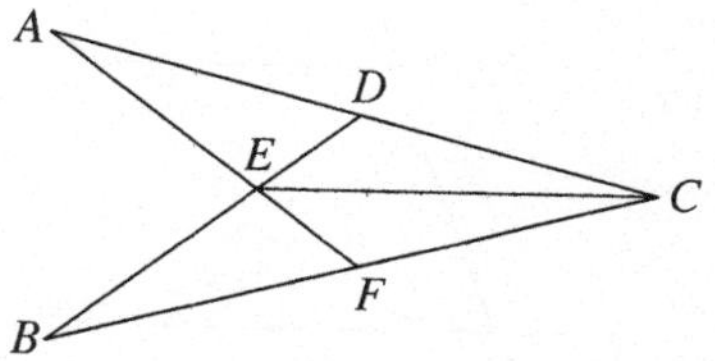


**17** First suppose that $\measuredangle C$ is right in $\triangle ABC$. Produce $\overline{BC}$ to $D$. We see that $\measuredangle ACB \cong \measuredangle ACD$. Since $\measuredangle BAC$ and $\measuredangle CBA$ are internal and opposite to external $\measuredangle ACD$, we conclude by Theorem 5.3.9 that

$$m\measuredangle ACB > m\measuredangle BAC \text{ and } m\measuredangle ACB > m\measuredangle CBA.$$

If, on the other hand, $\measuredangle C$ is obtuse, then the result still follows because $m\measuredangle BCA > m\measuredangle ACD$.

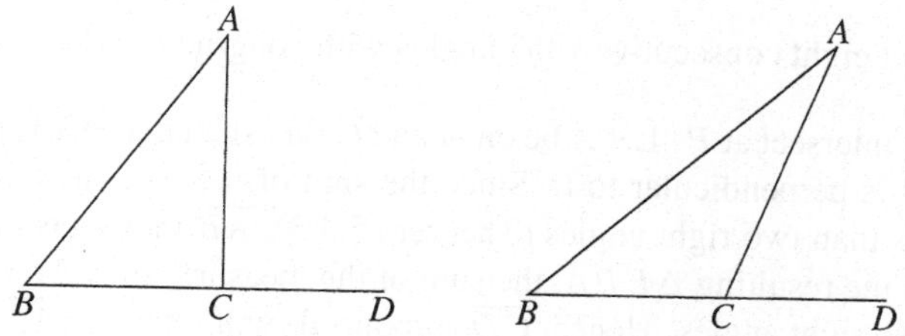


**19** Since the greater side is opposite the greater angle (Theorem 5.3.10), in $\triangle ABD$ we have $AD > BD$. Similarly, $BD > CD$. Therefore, $AD > CD$.

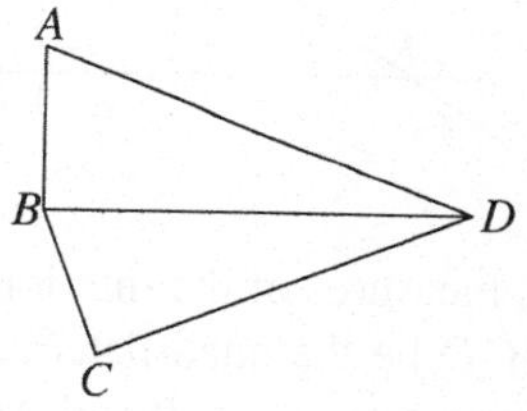


## 5.4 PARALLEL LINES

**1** No. If $ABCDE$ and $ABFGH$ are both regular pentagons, then $AEDCBFGH$ is obtained by joining the pentagons at $\overline{AB}$ and then removing this segment. The resulting figure is not convex.

**3(a)** Given $\overline{AB}$ and $P$ not on $\overline{AB}$, find $C$ on $\overline{AB}$ and join $\overline{CP}$. Find $D$ so that it lies on the same side of $\overline{CP}$ as $B$ and so that $\measuredangle CPD \cong \measuredangle ACP$. Use Construction 5.2.14 to accomplish this. Then $\overline{AB}$ and $\overline{PD}$ are parallel (Theorem 5.4.1).

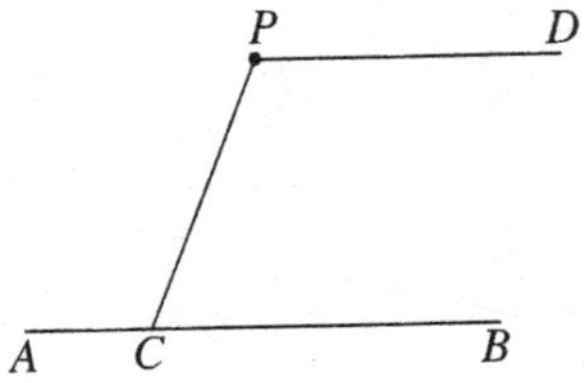


**3(c)** Construct an equilateral triangle and bisect one of its angles. This gives a 30° angle. Bisect this angle to obtain a 15° angle. Construct two copies of the 30° angle and one copy of the 15° angle, all adjacent to each other and sharing the same vertex. This gives the 75° angle. Another copy of the 30° angle will yield the 105° degree angle.

**3(e)** Construct perpendicular lines (Construction 5.2.11) $\overline{AB}$ and $\overline{AD}$, making sure that $\overline{AB} \cong \overline{AD}$. At $B$ and $D$ construct two more perpendicular lines meeting at $C$ and forming square $ABCD$.

**3(h)** Construct eight consecutive 135° angles with congruent sides.

**5** Let $l$ and $m$ intersect at $P$. Let $A$ be on $m$ and $C$ on $l$ such that $\overline{AB}$ is perpendicular to $m$ and $\overline{CD}$ is perpendicular to $l$. Since the sum of the measures of $\measuredangle CPA$ and $\measuredangle PAB$ is less than two right angles (Theorem 5.4.3), $\overline{AB}$ intersects $l$ at $E$ (Parallel Postulate). In the resulting $\triangle EPA$, the sum of the measures of $\measuredangle ECD$ and $\measuredangle CEB$ is less than two right angles. Hence, $\overline{CD}$ (produced) will intersect $\overline{AB}$ (produced).

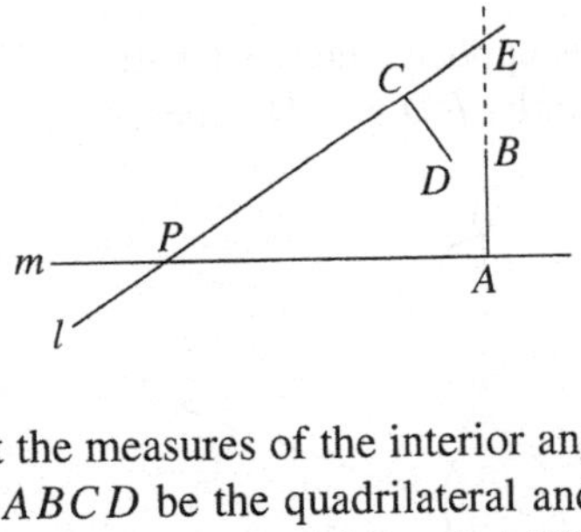


**7** We will prove here that the measures of the interior angles of a quadrilateral sum to four right angles. Let $ABCD$ be the quadrilateral and assume that it is convex. Join opposite vertices $A$ and $C$ to form $\triangle ACB$ and $ACD$. We know that

$$\mathrm{m}\measuredangle ABC + \mathrm{m}\measuredangle BCA + \mathrm{m}\measuredangle CAB = \text{ two right angles}$$

and

$$\mathrm{m}\measuredangle ADC + \mathrm{m}\measuredangle DCA + \mathrm{m}\measuredangle CAD = \text{ two right angles}$$

by Theorem 5.4.3. Since

$$\mathrm{m}\measuredangle DAB = \mathrm{m}\measuredangle CAD + \mathrm{m}\measuredangle CAB$$

and

$$\mathrm{m}\measuredangle DCB = \mathrm{m}\measuredangle DCA + \mathrm{m}\measuredangle BCA,$$

we conclude that

$$m\sphericalangle DAB + m\sphericalangle DCB + m\sphericalangle ABC + m\sphericalangle ADC = \text{ four right angles.}$$

If it is not convex, the same argument can be used, but we must be more careful as to how the two triangles are formed from the quadrilateral.

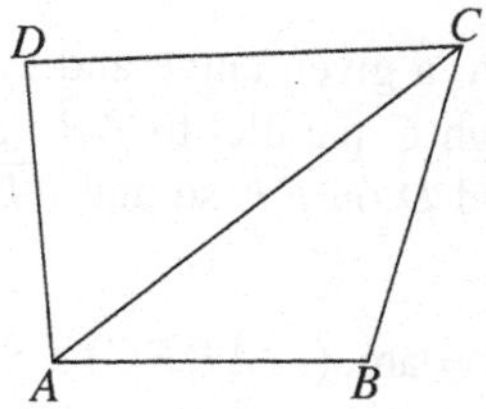


**9** This follows from Theorem 5.4.3.

**11** Given $\square ABCD$ as in the diagram, extend $\overline{DC}$ to $F$ so that $\overline{AF}$ is perpendicular to $\overline{DC}$ and extend $\overline{AB}$ to $E$ so that $\overline{CE}$ is perpendicular to $\overline{AB}$. Since $\overline{AD} \cong \overline{BC}$ and $\sphericalangle ADC \cong \sphericalangle ABC$ (Theorem 5.4.6) and $\sphericalangle FDA \cong \sphericalangle CBE$ (angle subtraction), $\Delta FDA \cong \Delta EBC$ by $AAS$. Thus, $\overline{FA} \cong \overline{CE}$. Also,

$$FC = FD + DC$$

and

$$AE = AB + BE,$$

so $\overline{FC} \cong \overline{AE}$. Lastly, $\sphericalangle FAE$ and $\sphericalangle FCE$ are right, so $FCEA$ is a rectangle. Its area is $FA \cdot AE$. Since the combined areas of $\Delta FDA$ and $\Delta EBC$ equal $FA \cdot BE$, the area of the parallelogram is

$$FA \cdot AE - FA \cdot BE = FA(AB + BE) - FA \cdot BE = FA \cdot AB.$$

(*Note:* The formula for the area of a parallelogram can also be obtained by joining $\overline{AC}$ and using the areas of the resulting triangles, but this would not be in line with the instructions of the exercise.)

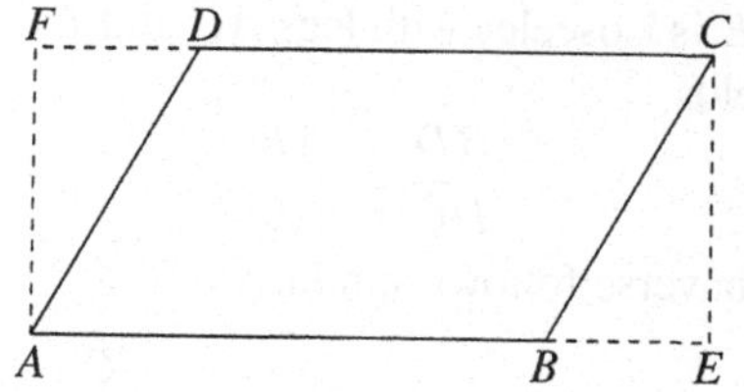


**13(a)** Let $\overline{AF}$ and $\overline{DG}$ be parallel. Let parallelograms $ABCD$ and $EFGH$ have congruent bases on $\overline{AF}$, namely $\overline{AB} \cong \overline{EF}$. Join $\overline{AH}$ and $\overline{BG}$ to form $\square ABHG$. (Why is $ABHG$ a parallelogram?) Then by Theorem 5.4.7,

$$\text{area}(\square ABCD) = \text{area}(\square ABGH) = \text{area}(\square EFGH).$$

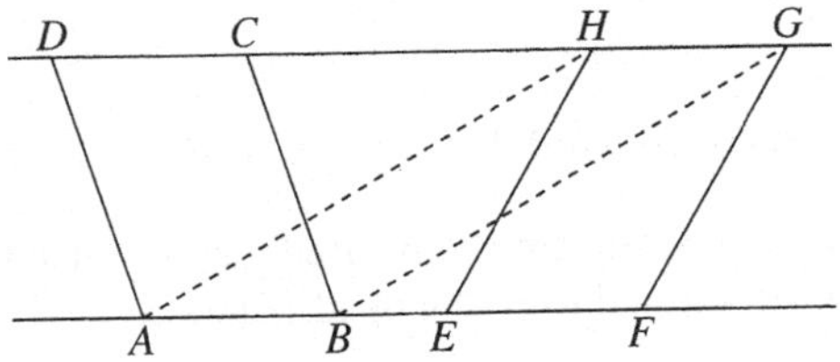


**15** Let $\sphericalangle EAB$ be congruent to a given angle and $\triangle ABC$ be congruent to a given triangle. Construct $\overline{EF}$ through $C$ parallel to $\overline{AB}$. Assume that $F$ is chosen such that $\overline{BF}$ is parallel to $\overline{AC}$. Find $D$ on $\overline{EF}$ so that $\overline{AE}$ is parallel to $\overline{BD}$. Therefore, by Theorem 5.4.7,

$$\text{area}(\square ABDE) = \text{area}(\square ABFC) = 2 \cdot \text{area}(\triangle ABC).$$

Now construct the midpoint $M$ of $\overline{AE}$ and the midpoint $N$ of $\overline{BD}$. We see that

$$\text{area}(\square ABDE) = 2 \cdot \text{area}(\square ABNM).$$

Therefore,

$$\text{area}(\square ABNM) = \text{area}(\triangle ABC).$$

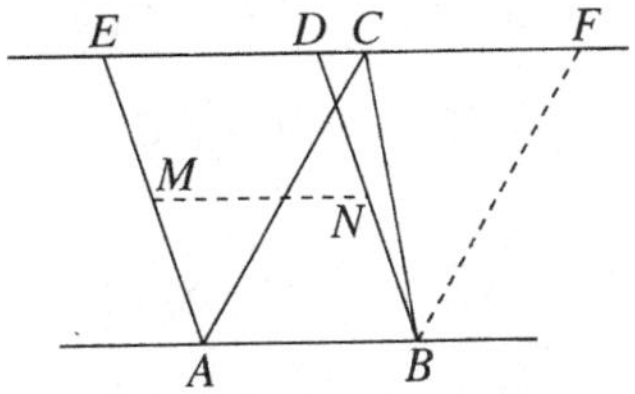


**17** Let $l$ and $m$ be two parallel lines. Take $A$ and $B$ on $m$. Find points $C$ and $D$ on $l$ such that both $\overline{AC}$ and $\overline{BD}$ are perpendicular to $l$. We conclude that $\overline{AC}$ is parallel to $\overline{BD}$ (Theorem 5.4.2). This means that $ACDB$ is a parallelogram and that $AC = BD$ (Theorem 5.4.6), and the distance $d$ between the two lines is $AC$.

**19** Suppose that $D$ is on $\overline{BC}$ in $\triangle ABC$ so that $\overline{DA}$ bisects $\sphericalangle BAC$. Construct $E$ so that it is on $\overline{AB}$ produced, yet $\overline{EC}$ is parallel to $\overline{AD}$. By Theorem 5.4.2, $\sphericalangle ECA \cong \sphericalangle CAD$. Similarly, $\sphericalangle CEA \cong \sphericalangle DAB$. Therefore, $\sphericalangle ECA \cong \sphericalangle AEC$, which means that $\triangle ACE$ is isosceles with legs $\overline{AC}$ and $\overline{EA}$. Since $\overline{AD}$ is parallel to $\overline{EC}$, Theorem 5.4.11 yields

$$\frac{BD}{DC} = \frac{AB}{AC}$$

since $EA = AC$. The converse follows similarly.

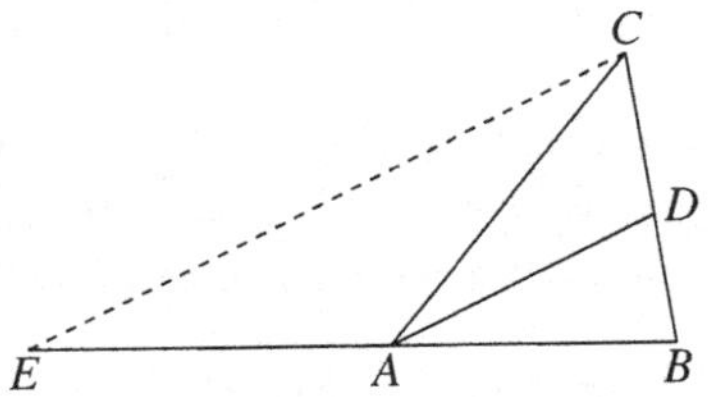

**21** We are given $\overline{DF}$ and $B$ on $\overline{AC}$. Find any $G$ such that $\overline{BG} \cong \overline{BC}$. Join $\overline{AG}$ and $\overline{CG}$. At $D$ copy $\sphericalangle GAC$ and at $F$ copy $\sphericalangle GCA$ so that the angles have $\overline{DF}$ as a common side and $H$ is the intersection of their other sides. Find $E$ on $\overline{DF}$ so that $\sphericalangle AGB \cong \sphericalangle DHE$. Hence, $\triangle ABG \sim \triangle DEH$, so

$$\frac{AB}{BC} = \frac{AB}{BG} = \frac{DE}{EH}.$$

Since $\sphericalangle GBC \cong \sphericalangle HEF$, $\triangle GBC \sim \triangle HEF$. Therefore, $\triangle HEF$ is isosceles with $\overline{EH} \cong \overline{EF}$. Therefore,

$$\frac{AB}{BC} = \frac{DE}{EF}.$$

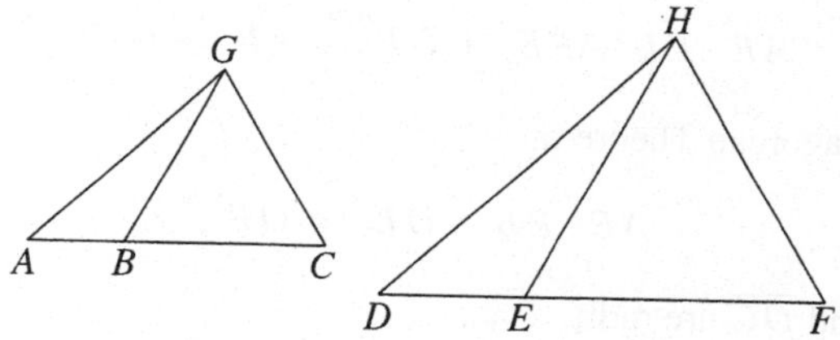


**23** Since congruence of line segments and angles are equivalence relations, similarity of polygons is an equivalence relation.

**25** We are given $\overline{AB}$ and $\overline{CD}$. Draw $\sphericalangle HGE$ such that $\overline{GH} \cong \overline{AB}$ and $\overline{GE} \cong \overline{CD}$. Join $\overline{HE}$ and extend $\overline{GH}$ to $J$ so that $\overline{HJ} \cong \overline{GE}$. Let $F$ be chosen so that $\overline{JF}$ is parallel to $\overline{HE}$ and the $\overline{GE}$ produced contains $F$. Then, by Theorem 5.4.11,

$$\frac{AB}{CD} = \frac{GH}{GE} = \frac{GH}{HJ} = \frac{GE}{EF} = \frac{CD}{EF}.$$

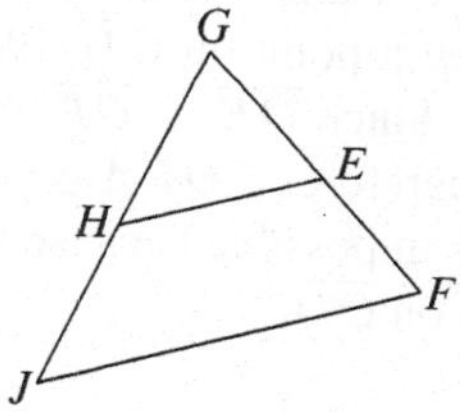


## 5.5 CIRCLES

**1(a)** From the diagram, if we let $x = AF = FB$ and $y = FE$, then the equation

$$AE \cdot EB + FE^2 = FB^2$$

translates to

$$(x + y)(x - y) + y^2 = x^2.$$

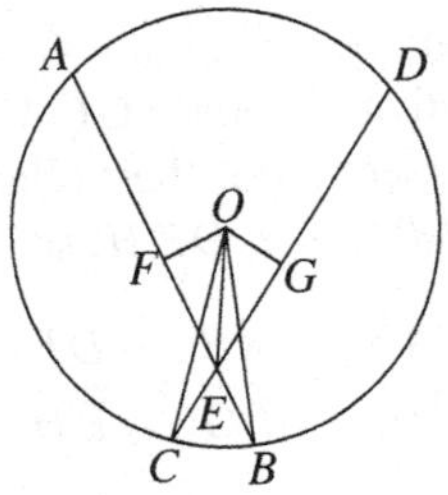


**2** We continue the proof of the Chord Theorem from the point after Equation 5.1 using the diagram from the solution to Exercise 1(a). Adding $OF^2$ to both sides of Equation 5.1 gives

$$AE \cdot EB + FE^2 + OF^2 = FB^2 + OF^2.$$

Thus, by the Pythagorean Theorem,

$$AE \cdot EB + OE^2 = OB^2.$$

Since both $\overline{OB}$ and $\overline{OC}$ are radii,

$$AE \cdot EB + OE^2 = OC^2.$$

A similar argument will show that

$$CE \cdot ED + OE^2 = OC^2.$$

Hence, $AE \cdot EB = CE \cdot ED$.

**5** Let $\overline{CD}$ be the perpendicular bisector of $\overline{AB}$ intersecting it at $E$. Assume that both lines are secants to the same circle. Suppose that center $O$ does not lie on $\overline{CD}$. First note that this implies that $O$ cannot be on $\overline{AB}$ because if it was, $O$ would be the midpoint of $\overline{AB}$ and thus be a point on $\overline{CD}$. We can now form two triangles by joining $\overline{OA}$, $\overline{OE}$, and $\overline{OB}$. Since $\overline{OE} \cong \overline{OE}$ and both $\overline{OA}$ and $\overline{OB}$ are radii, $\triangle OEA \cong \triangle OEB$ by SSS. Therefore, $\sphericalangle OEA \cong \sphericalangle OEB$, which means that $\overline{OE}$ is perpendicular to $\overline{AB}$. This is impossible because $\overline{CD}$ is already perpendicular to $\overline{AB}$. Hence, the center must lie on $\overline{CD}$.

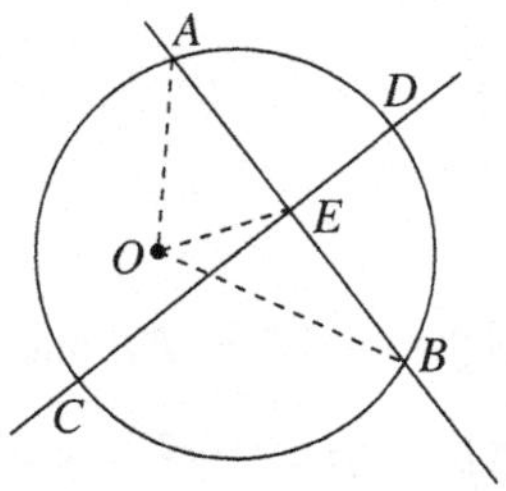


**7** The only way that two circles could be tangent and share the same center is when the one circle is on the interior of the other. Let this be the case and suppose that $O$

is the common center of the two circles. Let $A$ be the point of tangency. This means that $\overline{OA}$ is a radius of both circles. Find $B$ on the inner circle that does not lie on the outer circle and produce $\overline{OB}$ to $C$ on the outer circle. Then, $\overline{OC}$ is a radius of the outer circle, but $OC > OA$, a contradiction. Therefore, the two circles cannot share a common center.

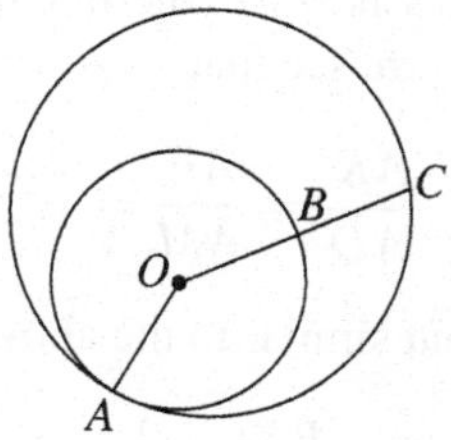


**9** The argument is similar to the solution to Exercise 7.

**11** By the proof of Theorem 5.5.8 we know that any two of the perpendicular bisectors of the sides of $\triangle ABC$ are not parallel. Let $\overline{OD}$ be the perpendicular bisector of $\overline{AB}$ and $\overline{OE}$ be the perpendicular bisector of $\overline{AC}$. Also by the proof of Theorem 5.5.8, we know that $\overline{OB} \cong \overline{OC}$. Therefore, $\sphericalangle OBC \cong \sphericalangle OCB$. Drop a perpendicular from $O$ to $F$ on $\overline{BC}$. Then, $\triangle BFO \cong \triangle CFO$ by AAS. Hence, $\overline{OF}$ is the perpendicular bisector of $\overline{BC}$.

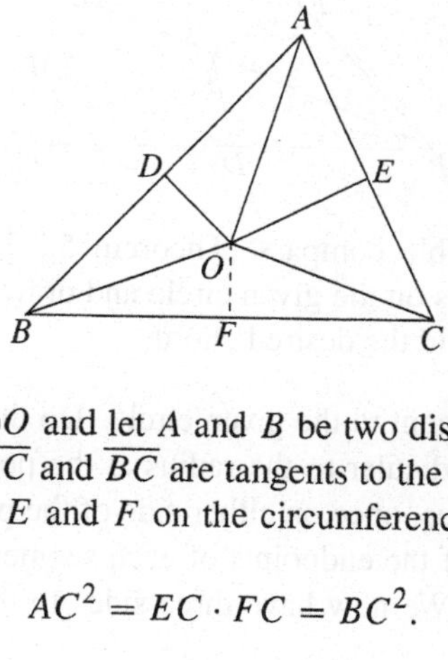


**13** Let $C$ be external to $\odot O$ and let $A$ and $B$ be two distinct points on the circumference of $\odot O$ such that $\overline{AC}$ and $\overline{BC}$ are tangents to the circle. Let $\overline{EF}$ be a secant of $\odot O$ containing $C$ with $E$ and $F$ on the circumference. By the Tangent-Secant Theorem (5.5.5),

$$AC^2 = EC \cdot FC = BC^2.$$

Hence, $\overline{AC} \cong \overline{BC}$.

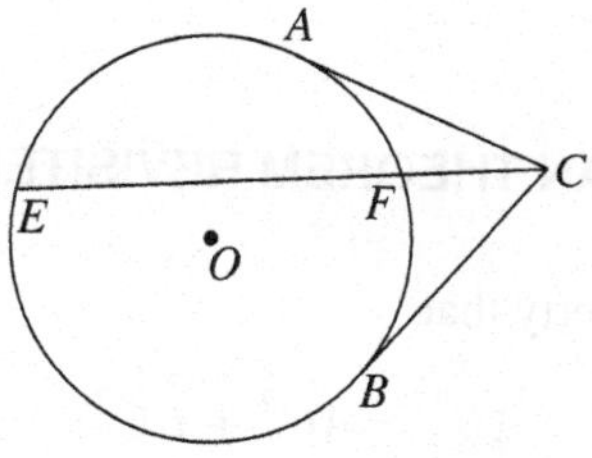

**15** In $\triangle ABC$, let $D$ be the midpoint of $\overline{BC}$, $E$ the midpoint of $\overline{AC}$, and $F$ the midpoint of $\overline{AB}$. Join $\overline{AD}$, $\overline{BE}$, and $\overline{CF}$. These are the medians of the triangle. Let $K$ be the intersection of $\overline{AD}$ and $\overline{BE}$. Let $N$ be the midpoint of $\overline{AE}$ and $M$ the midpoint of $\overline{EC}$. Draw lines parallel to $\overline{BE}$ through $N$ and $M$. Since $N$ is the midpoint of $\overline{AE}$, the line drawn through $N$ will intersect $\overline{AB}$ in its midpoint $F$ (Theorem 5.4.11 with $\triangle ABE$). Likewise, the line drawn through $M$ will intersect $\overline{BC}$ at $D$. Considering $\triangle ADM$, we see that

$$\frac{AK}{AD} = \frac{AE}{AM} = \frac{2}{3}$$

by Theorem 5.4.11. An argument similar to the above shows that

$$\frac{BK}{BE} = \frac{2}{3}.$$

This means that $\overline{AK}$ and $\overline{BE}$ intersect at a point cutting each segment two-thirds of their length from the vertex. Since the argument above works for any two pairs of medians, all three must be concurrent.

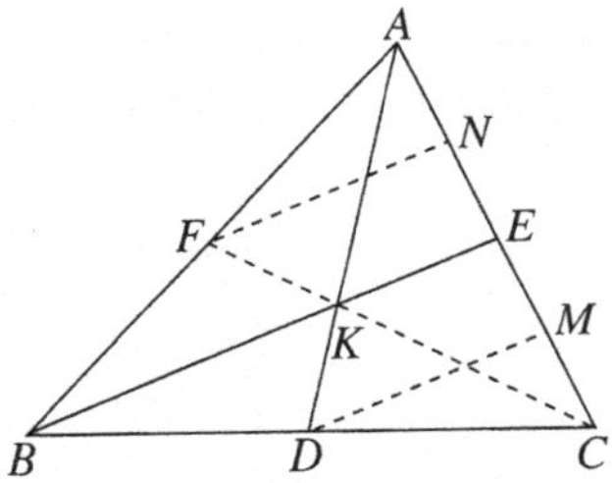


**16(a)** Using a noncollapsible compass (Theorem 5.2.7), measure the given line segment. Place the compass on the given circle and draw a new circle. Join the two points of intersection to yield the desired chord.

**16(d)** First construct a tangent to the given circle. Do this by drawing a radius and then constructing a perpendicular to the radius at the point of tangency. Copy the radius onto the tangent twice, once on either side of the point of tangency, using the point of tangency as one of the endpoints of each segment. At the other endpoints, construct perpendiculars. We now have three sides to the square. The last side is constructed similarly.

**16(g)** See Section 4.4.

## 5.6 THE PYTHAGOREAN THEOREM REVISITED

**1** Let $\triangle ABC$ have the property that

$$AB^2 = AC^2 + BC^2.$$

We want to show than $\sphericalangle ACB$ is right. Construct $\overline{CD}$ perpendicular to $\overline{AC}$ on the opposite side of $B$ such that $\overline{BC} \cong \overline{CD}$. (Note that at this point we do not know that $B$, $C$, and $D$ are collinear.) By Theorem 5.6.1,

$$AD^2 = AC^2 + CD^2.$$

By construction we then have

$$AD^2 = AC^2 + BC^2.$$

Hence, $\overline{AB} \cong \overline{AD}$, which yields $\triangle ABC \cong \triangle ADC$ by SSS. Therefore, $\sphericalangle ACB$ is right because it is congruent to $\sphericalangle ACD$.

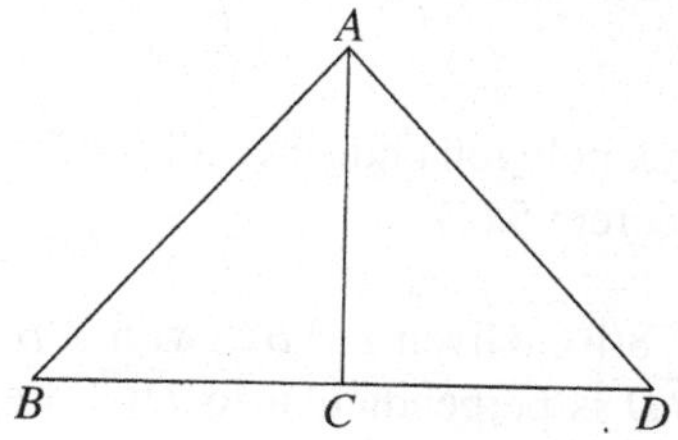


**3** The area of a rectangle is of the form $xy$, where $x,\ y > 0$. To construct a square with this area, we must construct a segment of length $\sqrt{xy}$. The desired square is then found by constructing perpendicular lines and copying line segments. We begin by constructing a segment of length $xy$. (Compare the following with Section 8.3.) Suppose that $y > 1$ (the $y < 1$ case is handled similarly). Let $\triangle ABC$ be drawn with $AB = x$ and $AC = 1$. Produce $\overline{AC}$ to $E$ so that $AE = y$. Copy $\sphericalangle ACB$ at $E$ to form $\sphericalangle AED$ with $D$ on $\overline{AB}$ produced. We see that $\triangle ABC \sim \triangle ADE$. Hence,

$$\frac{x}{1} = \frac{AD}{y}.$$

So, $AD = xy$. Constructing the square root of $xy$ is the same as constructing the mean proportional between 1 and $xy$. See page 112 of the text.

**5** To see that area$(\triangle ABG)$ = area$(\triangle DEF)$, we first note that we already have by Equation 5.5,

$$\frac{AB}{DE} = \frac{BC}{EF} = \frac{EF}{BG}.$$

Now drop two perpendiculars, one from $A$ to $\overline{BC}$ intersecting at $H$ and another from $D$ to $\overline{EF}$ at $I$. Therefore,

$$\text{area}(\triangle ABG) = \frac{BG \cdot AH}{2}$$

and

$$\text{area}(\triangle DEF) = \frac{EF \cdot DI}{2}.$$

However, in the proof of Theorem 5.6.2 we have that $\triangle ABH$ is similar to $\triangle DEI$, so

$$\frac{EF}{BG} = \frac{AB}{DE} = \frac{AH}{DI}.$$

Hence, $BG \cdot AH = EF \cdot DI$.

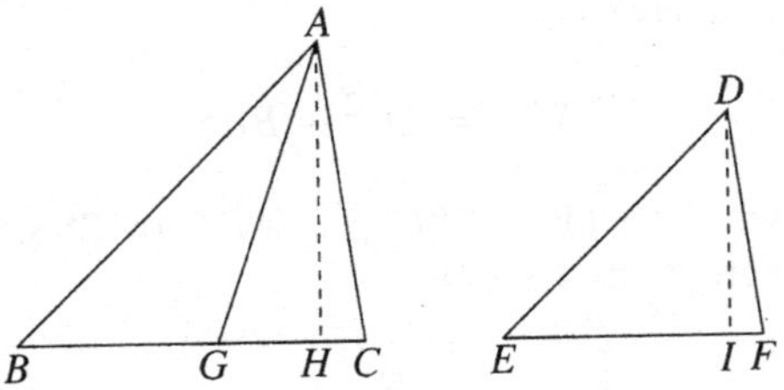


**7** Since a square is a convex polygon and since all squares are similar to each other, Theorem 5.6.4 implies Theorem 5.6.1.

**9** We will check Exercise 8(b). Given $\triangle ABC$ with $\measuredangle BAC$ obtuse, find $D$ on the $\overline{AC}$ produced such that $\overline{BD}$ is perpendicular to $\overline{DC}$. Let $a = AB$, $b = AC$, and $c = BC$. Let $\theta = \mathrm{m}\measuredangle BAC$. Therefore,

$$BC^2 = BA^2 + AC^2 + 2AC \cdot DA$$

translates to

$$c = a^2 + b^2 + 2ab\cos\theta$$

because

$$2ab\cos\theta = 2AB \cdot AC \cdot \frac{DA}{AB} = 2AC \cdot DA.$$

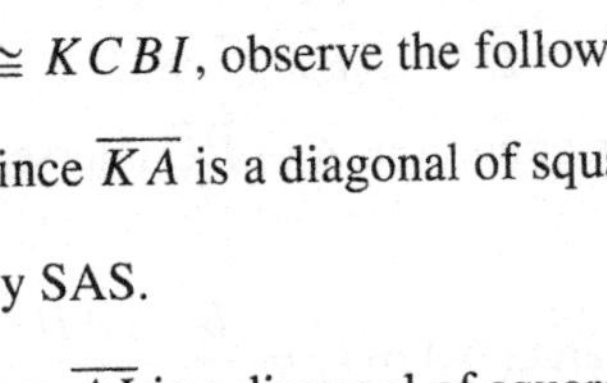


**11(a)** To see that $KGHI \cong KCBI$, observe the following:

- $\triangle KAG \cong \triangle KAC$ since $\overline{KA}$ is a diagonal of square $KCAG$.
- $\triangle GAH \cong \triangle CAB$ by SAS.
- $\triangle AIH \cong \triangle AIB$ since $\overline{AI}$ is a diagonal of square $ABIH$.

Since corresponding parts of $KGHI$ are congruent to corresponding parts of $KCBI$, it is natural to say that these two figures are congruent.

**11(c)** By Exercise 11(b),

$$\text{area}(CEDB) + \text{area}(CBA) + \text{area}(DFE)$$

equals

$$\text{area}(KCAG) + \text{area}(ABIH) + \text{area}(CBA) + \text{area}(GAH).$$

Since $\triangle DFE \cong \triangle GAH$,

$$\text{area}(CEDB) = \text{area}(KCAG) + \text{area}(ABIH).$$

# CHAPTER 6

# ARCHIMEDES

## 6.1 THE ARCHIMEDEAN LIBRARY

**1** From calculus, the center of mass of a planar lamina is $(\bar{x}, \bar{y})$, where

$$\bar{x} = \frac{M_y}{M}, \quad \bar{y} = \frac{M_x}{M},$$

$M_x$ is the moment about the $x$-axis, $M_y$ is the moment about the $y$-axis, and $M$ is the mass (area) of the lamina. To answer the questions, take a section that is described by a parabola with vertex at the origin and that passes through the point $(a, b)$. The equation for this parabola is

$$x^2 = \frac{a^2}{b}y.$$

Now compute $M$, $M_y$, and $M_x$:

$$M = \int_{-a}^{a} \left(b - \frac{b}{a^2}x^2\right) dx = \frac{4}{3}ab,$$

$$M_y = \int_{-a}^{a} x\left(b - \frac{b}{a^2}x^2\right) dx = 0,$$

and

$$M_x = \int_{-a}^{a} \frac{1}{2}\left(b^2 - \frac{b^2}{a^4}x^4\right) dx = \frac{4}{5}ab^2.$$

Therefore,

$$\bar{x} = 0,$$

showing that the section's center of gravity is on its axis of symmetry. Furthermore,

$$\bar{y} = \frac{3}{5}b.$$

Letting $\overline{OP}$ be the axis with $P$ on the $x$-axis, $O$ on the line $y = b$, and $G$ the center of gravity, we see that

$$\frac{PG}{GO} = \frac{\frac{3}{5}b}{\frac{2}{5}b} = \frac{3}{2}.$$

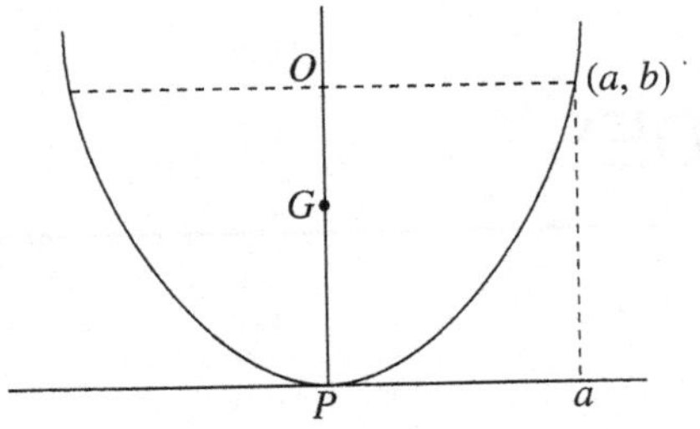


**3(a)** The area enclosed by the ellipse is

$$\frac{4b}{a}\int_0^a \sqrt{a^2 - x^2}\, dx.$$

Using the trigonometric substitution $x = a\sin\theta$, the integral is equivalent to

$$4ab\int_0^{\pi/2}\left(\frac{1}{2} + \frac{1}{2}\cos 2\theta\right) d\theta = \pi ab.$$

**5(a)** Integrate:

$$\int_0^{2\pi} \frac{1}{2}a^2\theta^2\, d\theta = \left.\frac{a^2\theta^3}{6}\right]_0^{2\pi} = \frac{4}{3}\pi^3 a^2.$$

**7** To find the volume of revolution of $x^2 = r^2y/h$ from $y = 0$ to $y = h$ using horizontal disks, first note that the volume of one of the disks is

$$\Delta V = \pi\left(\sqrt{r^2y/h}\right)^2 \Delta y.$$

Therefore,

$$V = \int_0^h \frac{\pi r^2}{h} y\,dy = \frac{\pi r^2}{h}\left(\frac{y^2}{2}\right)\Bigg]_0^h = \frac{1}{2}\pi r^2 h.$$

**9(a)** Let $A_1$ be the area of the semicircle on $\overline{AC}$, $A_2$ be the area of the semicircle on $\overline{AB}$, and $A_3$ be the area of the semicircle on $\overline{BC}$. Then

$$A_1 = \frac{\pi AC^2}{8}, \ A_2 = \frac{\pi AB^2}{8}, \ A_3 = \frac{\pi BC^2}{8}.$$

When $\overline{AD}$ and $\overline{DC}$ are joined, we find that the area of the shaded region is

$$\begin{aligned} A_1 - A_2 - A_3 &= \frac{\pi}{8}(AC^2 - AB^2 - BC^2) \\ &= \frac{\pi}{8}(AC^2 - [AD^2 - DB^2] - [DC^2 - DB^2]) \\ &= \frac{\pi}{8}(AC^2 - AD^2 - DC^2 + 2DB^2) \\ &= \frac{\pi DB^2}{4}. \end{aligned}$$

The second and last equalities hold by the Pythagorean Theorem.

## 6.2 THE METHOD OF EXHAUSTION

**1** In the diagram each successive right triangle is constructed on the hypotenuse of the previous one. If the smallest triangle has legs of length 1, the largest has a hypotenuse of length $\sqrt{17}$. If constructed, the next triangle would overlap the first.

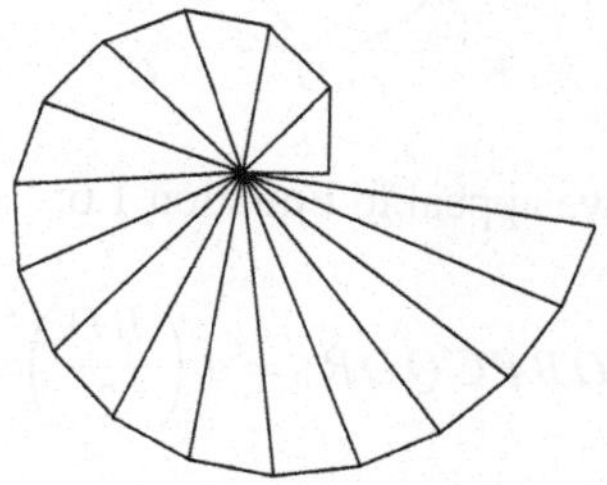

**3** Let $AB < DE$ and $DE = 4CC'$. Find points $F_1,\ F_2,\ F_3$ on $\overline{DE}$ such that each is congruent to $\overline{CC'}$ and

$$DE = DF_3 + F_3F_2 + F_2F_1 + F_1E.$$

- Find $H_1$ on $\overline{AB}$ such that $AH_1 < \frac{1}{2}AB$. Then

$$DF_1 > \frac{1}{2}DE > \frac{1}{2}AB > AH_1.$$

- Next, find $H_2$ on $AH_1$ such that $AH_2 < \frac{1}{2}AH_1$. Then
$$DF_2 > \frac{1}{2}DF_1 > \frac{1}{2}AH_1 > AH_2.$$
- Finally, take $H_3$ on $AH_2$ such that $AH_3 < \frac{1}{2}AH_2$. As before we have
$$DF_3 > AH_3,$$
but this time $\overline{DF_3} \cong \overline{CC'}$. This means that $AH_3$ is the sought-after segment of $\overline{AB}$.

This is just the same argument as in the text. It can be generalized to any even number of segments.

**5** To prove Equation 6.9, let $\Box EFGH$ be inscribed within $\odot O$ and let $\Box E'F'G'H'$ be circumscribed about it. Then,

$$\text{area}(\Box E'F'G'H') = 4OF^2.$$

and because $EF = \sqrt{2}OF$,

$$\text{area}(\Box EFGH) = 2OF^2.$$

So,

$$2\ \text{area}(\Box EFGH) = \text{area}(\Box E'F'G'H').$$

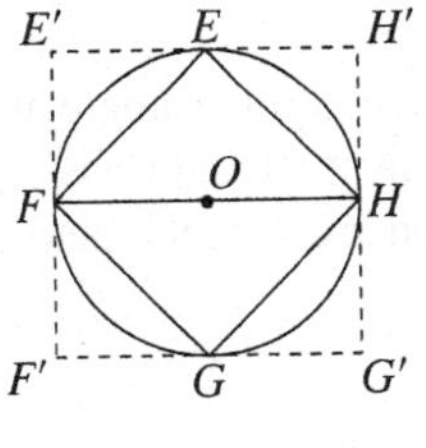


To prove Equation 6.11 we appeal to Equation 1.6:

$$\text{area}(AOBPCQDR) = 4\left(\frac{BD}{2}\right)^2 \sin\frac{2\pi}{8}$$

and

$$\text{area}(EKFLGMHNE) = 4\left(\frac{FH}{2}\right)^2 \sin\frac{2\pi}{8}.$$

Hence, the ratio of these two areas is $BD^2/FH^2$.

**7** We use the notation from Theorem 6.2.4. Suppose that

$$\text{area}(\odot ABCD) < \text{area}(\Delta XYZ).$$

By Lemma 6.2.2 there exists a regular polygon of area $A_n$ that is circumscribed about $\odot ABCD$ such that

$$A_n - \text{area}(\odot ABCD) < \text{area}(\Delta XYZ) - \text{area}(\odot ABCD).$$

From this we conclude that

$$A_n < \text{area}(\Delta XYZ).$$

Now let $p$ be the perimeter of the circumscribed polygon. Because $p$ is greater than the circumference of the circle,

$$p > YZ.$$

Since

$$\text{area of a regular polygon} = \frac{\text{apothem} \times \text{perimeter}}{2},$$

we see that

$$A_n = \frac{XY \cdot p}{2} > \frac{XY \cdot YZ}{2} = \text{area}(\Delta XYZ),$$

a contradiction.

**9** We will take advantage of the hint and the results from Exercise 8:

$$\sqrt{3} = \frac{1}{15}\sqrt{26^2 - 1} > \frac{1}{15}\sqrt{\left(26 - \frac{1}{51}\right)^2} = \frac{1}{15}\left(26 - \frac{1}{51}\right) = \frac{1325}{765} = \frac{265}{153}.$$

**10(c)** Note that since $r = 1$, the area of the circle is $\pi$. To find the areas of the octagons, we will use Equation 1.6. First, the area of the inscribed octagon is

$$4 \sin \frac{\pi}{4} \approx 2.828.$$

Second, since the radius of the circumscribed octagon is $\sec \pi/8$, the area of the circumscribed octagon is

$$4 \sec^2 \frac{\pi}{8} \sin \frac{\pi}{4} \approx 3.314.$$

Thus, $2.828 < \pi < 3.314$.

**13** Using Archimedes' approximation of $\pi \approx 22/7$,

$$\frac{\pi r^2}{(2r)^2} = \frac{\pi}{4} \approx \frac{1}{4} \cdot \frac{22}{7} = \frac{11}{14}.$$

**15(a)** Let $V_1$ be the volume of one sphere of radius $r_1$ and let $V_2$ be the volume of the other sphere of radius $r_2$. The result then translates as

$$\frac{V_1}{V_2} = \frac{r_1}{r_2}.$$

## 6.3 THE METHOD

**1(a)** 30/7

**1(c)** $bx/(a+2b)$

**3** Let $A$, $B$, $C$, $D$, $E$, and $F$ be collinear and equally spaced. Assume that equal weights are placed with their centers of gravity at these six points. When we group $A$, $B$, and $C$ into a system and do the same thing for $D$, $E$, and $F$, we conclude by Theorem 6.3.2 that $B$ is the center of gravity for the first system and $E$ is the center of gravity for the second system. Therefore, we may view the weights as a single system with the center of gravity at the midpoint $M$ of $\overline{BE}$ (Theorem 6.3.2). Since

$$BM = \frac{BE}{2}$$

and

$$BE = BC + CD + DE = 3CD,$$

we have that

$$BM = \frac{3}{2}CD = CD + \frac{CD}{2} = BC + \frac{CD}{2},$$

from which follows

$$CM = BM - BC = \frac{CD}{2}.$$

Therefore, $M$ is the midpoint of $\overline{CD}$.

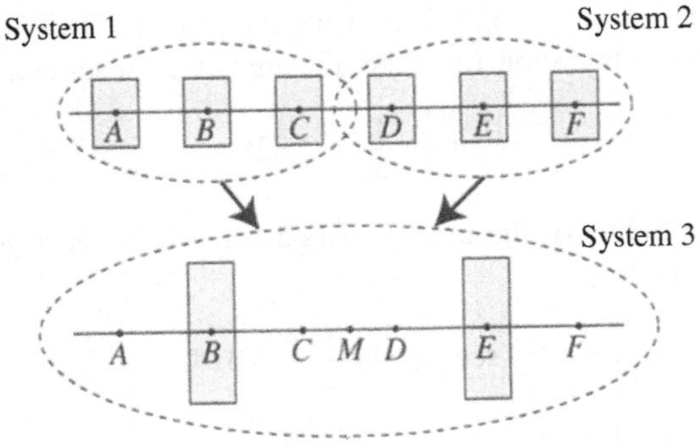


**5** Given $\square ABCD$ let $M$, $N$, $O$, and $P$ be the midpoints of $\overline{AB}$, $\overline{DC}$, $\overline{AD}$, and $\overline{BC}$, respectively. By Theorem 6.3.4, the center of gravity $G$ of $\square ABCD$ is on both $\overline{MN}$ and $\overline{OP}$. Join $\overline{DG}$ and $\overline{GB}$. We claim that $D$, $G$, and $B$ are collinear. Since $MBPG$ and $OGND$ are also parallelograms, $\overline{DN} \cong \overline{GP}$ and $\overline{GN} \cong \overline{BP}$. Since $\sphericalangle DNG \cong \sphericalangle GPB$, we conclude that

$$\triangle DNG \cong \triangle GPB$$

by SAS. Now, $\sphericalangle OGD \cong \sphericalangle BGP$ since they are vertical, and then because

$$m\sphericalangle OGD + m\sphericalangle DGN + m\sphericalangle NGP = \text{two right angles},$$

we have

$$m\sphericalangle BGP + m\sphericalangle DGN + m\sphericalangle NGP = \text{two right angles.}$$

Therefore, $D$, $G$, and $B$ are collinear by Theorem 5.3.8, and $G$ is on $\overline{DB}$. Similarly, $G$ is on $\overline{AC}$.

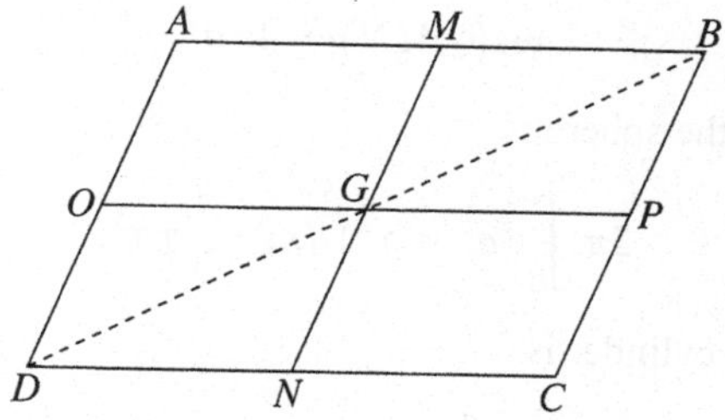


**7** To construct the center of gravity of a circle, we need to find its center. We use Euclid's proof (*Elements* III.1). Let $\overline{AB}$ be any chord of $\odot AB$. Let $M$ be the midpoint of $\overline{AB}$ and let chord $\overline{CD}$ contain $M$ and be perpendicular to $\overline{AB}$. We claim that the midpoint $O$ of $\overline{CD}$ is the center of the circle. Suppose that instead it is point $P$. We know that $P$ does not lie on $\overline{CD}$. We now join $\overline{AP}$, $\overline{MP}$, and $\overline{BP}$. Since $\overline{AP}$ and $\overline{BP}$ are radii, they are congruent. Hence, $\triangle AMP \cong \triangle BMP$ by SSS. Therefore, $\sphericalangle AMP \cong \sphericalangle BMP$, so these are both right angles, but this is impossible because $\sphericalangle AMO$ and $\sphericalangle BMO$ are right angles.

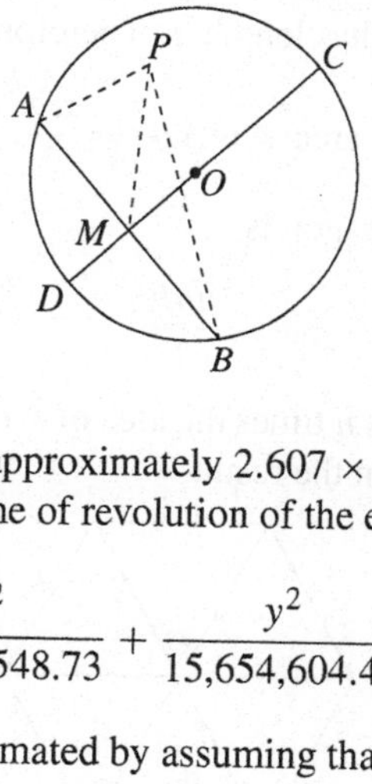


**9** The volume of the Earth is approximately $2.607 \times 10^{11}$ cubic miles. This can be confirmed by finding the volume of revolution of the ellipse:

$$\frac{x^2}{15{,}707{,}548.73} + \frac{y^2}{15{,}654{,}604.43} = 1.$$

(Of course, this can also be estimated by assuming that the Earth is a sphere.)

**11(a)** We first note that the equation for $\overline{AF}$ is

$$y = a - x,$$

the equation for $\odot ABCD$ is

$$x^2 + y^2 = a^2,$$

and $EF = 4a$. Therefore,

$$\text{area}(\odot OP) = \pi(a^2 - x^2),$$
$$\text{area}(\odot QR) = \pi(a - x)^2,$$

and

$$\text{area}(\odot MN) = 4\pi a^2.$$

**11(c)** The volume of the sphere is

$$2\pi \int_0^a (a^2 - x^2)\,dx = \frac{4}{3}\pi a^3,$$

and the volume of the cylinder is

$$8\pi \int_0^a a^2\,dy = 8a^3\pi.$$

**13(a)** Recall that a right-angled conoid is also known as a paraboloid of revolution. Let $h$ be the length of its axis. The center of gravity of the conoid is located on its axis $3h/5$ units from the vertex.

## 6.4 PRELIMINARIES TO THE PROOF

**1** Let $AB$ be a side of a regular convex $n$-gon with center $O$. Assume that each side is of length $b$ and the apothem has length $a$. Therefore,

$$\text{area}(\triangle OAB) = \frac{ab}{2}.$$

Hence, the area of the regular $n$-gon is

$$\frac{nab}{2}$$

because the area of the $n$-gon is $n$ times the area of $\triangle OAB$. Since $nb$ is the perimeter of the polygon, we have proven the result.

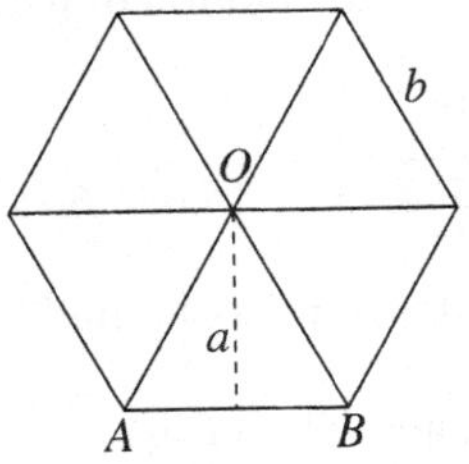


**2(a)** For the $n$-gon of the diagram, $\measuredangle AOB = 2\pi/n$. Let $a$ be the length of the apothem of the circumscribed $n$-gon. Since

$$AC = AO \cdot \sin\frac{\pi}{n}$$

and

$$AO = a \sec \frac{\pi}{n},$$

we conclude that

$$AB = 2a \tan \frac{\pi}{n}.$$

Since the apothem of the circumscribed $n$-gon is the radius of the inscribed $n$-gon,

$$DF = a \sin \frac{\pi}{n}.$$

Thus,

$$DE = 2a \sin \frac{\pi}{n}.$$

Therefore,

$$\frac{\text{side of } P_C}{\text{side of } P_I} = \frac{AB}{DE} = \frac{\tan \pi/n}{\sin \pi/n}.$$

We want this ratio to be less than 1.00005. The minimum $n$ that accomplishes this is $n = 6$.

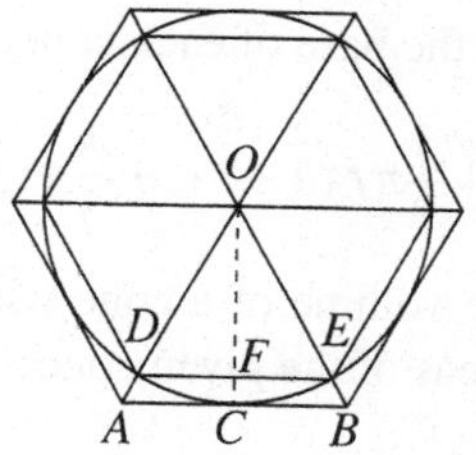


**3** Let $AB = a$. Construct $\overline{AC}$ perpendicular to $\overline{AB}$. Draw a circle with center $B$ and radius $c$. The circle intersects $\overline{AC}$ (produced) at some $D$. Then $\triangle ABD$ is the desired triangle.

**5** By Exercise 1,

$$\text{area}(P_C) = \frac{nAB \cdot OC}{2}$$

and

$$\text{area}(P_I) = \frac{nDE \cdot OF}{2}.$$

Furthermore, because $\triangle OAC \sim \triangle ODE$,

$$\frac{OC}{OF} = \frac{AB}{DE}.$$

Therefore,

$$\frac{\text{area}(P_C)}{\text{area}(P_I)} = \frac{AB \cdot OC}{DE \cdot OF} = \frac{(\text{side of } P_C)^2}{(\text{side of } P_I)^2}.$$

**7(a)** $\sqrt{65}$

**7(c)** $4\pi\sqrt{65} + 16\pi$

**9(a)** Let $V_1$ and $V_2$ be the volumes of cones with bases of radius $r$. Let $h_1$ be the height of the first cone and $h_2$ the height of the second cone. Then,

$$\frac{V_1}{V_2} = \frac{\pi r^2 h_1/3}{\pi r^2 h_2/3} = \frac{h_1}{h_2}.$$

**11(a)** $70\pi$

## 6.5 THE VOLUME OF A SPHERE

**1** Let $a$ be the length of the apothem and $s$ be the surface area of revolution of the standard polygon. Then,

$$\text{area(standard polygon)} = \frac{1}{3}as.$$

**3** Let $V_1, V_2, \ldots, V_n$ be the volumes of $n$ cones. Suppose that the height of each cone is $h$ and that the radius of the base of each cone is $r_n$. Then,

$$V_1 + V_2 + \cdots + V_n = \frac{1}{3}\pi r_1^2 h + \frac{1}{3}\pi r_2^2 h + \cdots + \frac{1}{3}\pi r_n^2 h = \frac{1}{3}\pi h(r_1 + r_2 + \cdots + r_n),$$

which is the expression for the volume of a cone with height $h$ and base with area equal to the sum of the base areas of the given cones.

**5** Let $V$ be the volume of a sphere of radius $r$. We note that the area of a great circle of this sphere equals $\pi r^2$. According to Theorem 6.5.8,

$$V = 4\left(\frac{1}{3} \cdot \pi r^2 \cdot r\right) = \frac{4}{3}\pi r^3.$$

This is equivalent to the expression of Corollary 6.5.9 as noted on page 226.

**7(a)** From the proof of Theorem 6.5.8, we should calculate $d$ and $e$ as arithmetic means, but which ones? We will choose $e$ as the arithmetic mean between $a$ and $b$ and then choose $d$ to be the arithmetic mean between $a$ and $e$. That is,

$$a - e = e - b \text{ and } a - d = d - e.$$

**7(b)** The solution is due to Eutocius. Let $a > d > e > b$, where $e$ is the arithmetic mean between $a$ and $b$ and $d$ is the arithmetic mean between $a$ and $e$. Begin by finding $x$ such that

$$\frac{d}{a} = \frac{x}{d}. \qquad (*)$$

Thus,

$$\frac{a-d}{a} = \frac{a}{a} - \frac{d}{a} = \frac{d}{d} - \frac{x}{d} = \frac{d-x}{d}.$$

Since $a > d$, by $(*)$ we have that

$$a - d > d - x,$$

and since $d$ is the arithmetic mean between $a$ and $e$,

$$d - e > d - x.$$

We note that this implies that $x > e$. Now find $y$ such that

$$\frac{d}{x} = \frac{x}{y}.$$

This implies that

$$d - x > x - y.$$

Therefore,

$$a - e > x - y,$$

and because $e$ is the arithmetic mean between $a$ and $b$,

$$e - b > x - y.$$

Note that because $x > e$, we also have $y > b$. Finally, because

$$a^2x^2 = d^4$$

and

$$dy = x^2,$$

we may then conclude that

$$\frac{a^3}{d^3} = \frac{a}{y} < \frac{a}{b}.$$

# CHAPTER 7

# FRANÇOIS VIÈTE

## 7.1 THE ANALYTIC ART

**1** When we write an equation such as $x^2 + 4x - 1 = 0$, we are assuming that $x$ is a solution to the equation. We then essentially work backwards to find its value. However, sometimes the steps are not reversible (like when we square both sides of the equation), so we might have to check that our derivation works as a solution.

**3** Since $a^n - b^{n-m}a^m = z^n$ and $b^{n-m}e^m - e^n = z^n$,

$$\begin{aligned} b^{n-m}e^m - e^n &= a^n - b^{n-m}a^m, \\ b^{n-m}e^m + b^{n-m}a^m &= a^n + e^n, \\ b^{n-m} &= \frac{a^n + e^n}{e^m + a^m}. \end{aligned}$$

Therefore, substituting into the first equation gives

$$a^n - \frac{a^m(e^n + a^n)}{e^m + a^m} = z^n.$$

*Revolutions of Geometry*. By Michael O'Leary

**5(a)** If $a^3 - 3b^2a = z^s$ and $a + b = e$, then

$$\begin{aligned} e^3 - 3be^2 &= (a+b)^3 - 3b(a+b)^2 \\ &= a^3 + 3a^2b + 3ab^2 + b^3 - 3a^2b - 6ab^2 - 3b^3 \\ &= a^3 - 3ab^2 - 2b^3 \\ &= z^s - 2b^3. \end{aligned}$$

**7** Let $c$ and $d$ be roots of $a^3 + z^s = 3b^2a$. (Note that in the exercise, $a_1 = c$ and $a_2 = d$.) This means that

$$c^3 + z^s = 3b^2c$$

and

$$d^3 + z^s = 3b^2d.$$

First,

$$\begin{aligned} 3b^2c - c^3 &= 3b^2d - d^3, \\ 3b^2c - 3b^2d &= c^3 - d^3, \\ 3b^2(c-d) &= c^3 - d^3, \\ 3b^2 &= \frac{c^3 - d^3}{c-d}, \\ 3b^2 &= c^2 + cd + d^2. \end{aligned}$$

Second, because

$$3b^2 = \frac{z^s + c^3}{c}$$

and

$$3b^2 = \frac{z^s + d^3}{d},$$

we have

$$\begin{aligned} z^sd + c^3d &= z^sc + d^3c, \\ z^sd - z^sc &= d^3c - c^3d, \\ z^2(d-c) &= dc(d^2 - c^2), \\ z^s &= \frac{dc(d^2 - c^2)}{d-c}, \\ z^s &= d^2c + dc^2. \end{aligned}$$

**9(a)** Let $\odot O$ be the unit circle. Since

- $\sin x = AE$,
- $\sin y = BC$,

- m$\measuredangle HAB = (x + y)/2$ (because $\measuredangle HAB$ is an inscribed angle for $\widehat{BF}$),
- m$\measuredangle AGB = (x - y)/2$ (because $\measuredangle AGB$ is an inscribed angle for $\widehat{AB}$),
- $GB = 2$, and
- $\measuredangle GAB$ is right (because $\overline{GB}$ is a diameter),

we conclude that

$$\begin{aligned}\sin x - \sin y &= AE - BC \\ &= AH \\ &= AB\cos\left(\frac{x+y}{2}\right) \\ &= 2\sin\left(\frac{x-y}{2}\right)\cos\left(\frac{x+y}{2}\right).\end{aligned}$$

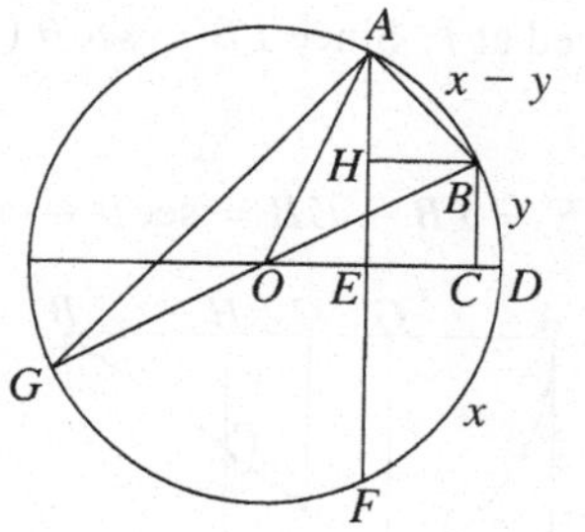


## 7.2 THREE PROBLEMS

**1** See Exercise 3 from Section 5.6. Its solution is in this manual.

**3** $x = y\cot\left(\frac{\pi y}{2a}\right)$

**5** Using a diagram similar to that from Exercise 4, divide $\overline{AB}$ into five congruent segments. Let the top one be $\overline{BM}$. Construct a perpendicular to $\overline{AB}$ at $M$ that intersects the quadratrix at $H$. Then $\measuredangle BAH$ is one fifth of a right angle.

**7** We see from the figure that

$$CP = DP + CD = a + b\sec(\pi/2 - \theta) = a + b\csc\theta.$$

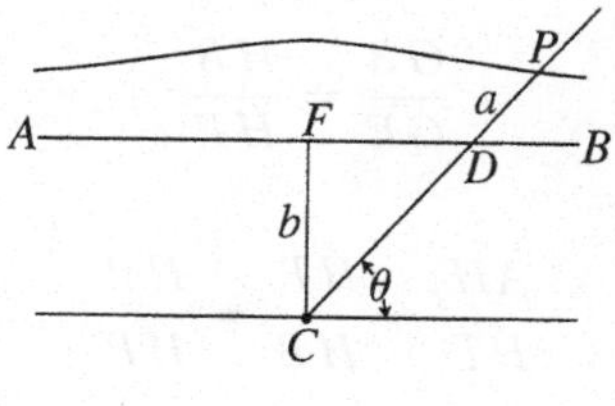

**9(a)** The conchoid is designed to transfer such distances, so a conchoid with pole $F$, ruler $\overline{CH}$, and distance $CF$ will enable us to find the required segment. Specifically, $I$ is the intersection of this conchoid with the $\overline{BC}$ produced. The figure illustrates this.

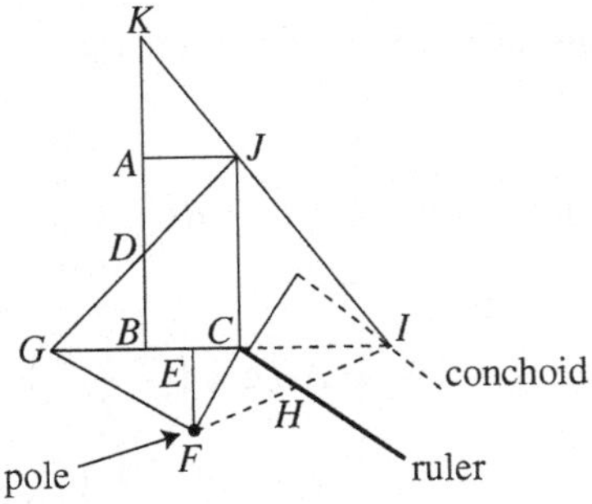


**11(a)** Let $r = PB$, $\theta = \text{m}\measuredangle ABP$, and $AB = 1$. Construct a perpendicular at $A$ that intersects the $\overline{PB}$ produced at $I$. Since $IB = \sec\theta$ (in $\triangle BIA$), $EB = \cos\theta$ (in $\triangle ABE$), and $\overline{IE} \cong \overline{PB}$,

$$r = IE = IB - EB = \sec\theta - \cos\theta.$$

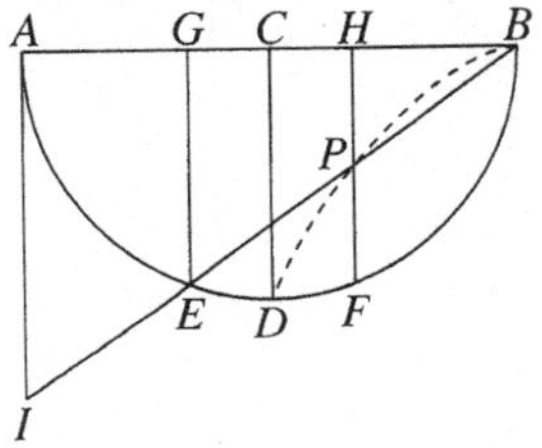


**11(d)** We will use the diagram from the solution to Exercise 11(a). In the cissoid, $\overline{GE} \cong \overline{HF}$ and $\overline{AG} \cong \overline{HB}$, so

$$\frac{GB}{GE} = \frac{AH}{HF}.$$

By the Chord Theorem (5.5.3),

$$\frac{AH}{HF} = \frac{HF}{HB}.$$

Finally, since $\triangle GBE \sim \triangle HBP$,

$$\frac{GB}{GE} = \frac{HB}{HP}.$$

Therefore,

$$\frac{AH}{HF} = \frac{HF}{HB} = \frac{HB}{HP}.$$

**13** Let $\measuredangle ABC$ be right in $\triangle ABC$. Choose any $D$ on the $\overline{BC}$ produced beyond $B$. This means that $DC \cdot AC$ can represent the area of an arbitrary rectangle, yet with area greater than the area of $\triangle ABC$. Construct $\overline{CE}$ parallel to $\overline{DA}$. Extend $\overline{BC}$ to a point $F$ at an arbitrary distance from $C$. By Postulate 7.2.2, there exists $G$ on $\overline{CF}$ such that $\overline{HG} \cong \overline{AC}$. (*Note:* With reference to the postulate, the angle is $\measuredangle ECF$ and the external point is $A$.) By Theorem 5.4.11,

$$\frac{DC}{AH} = \frac{CG}{HG} = \frac{CG}{AC}.$$

Hence, $\triangle ABG$ is the sought-after triangle because the difference in their bases is $CG$, the difference in their hypotenuse is $AH$, and $CG \cdot AH = DC \cdot AC$.

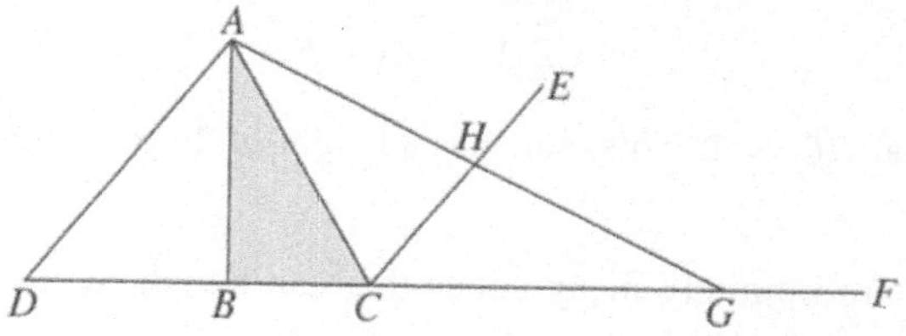


## 7.3 CONIC SECTIONS

**1(a)** The Chord Theorem (5.5.3) gives the result because $\overline{BC}$ is a diameter and $\overline{PN}$ is perpendicular to $\overline{BC}$ with $N$ on $\overline{BC}$.

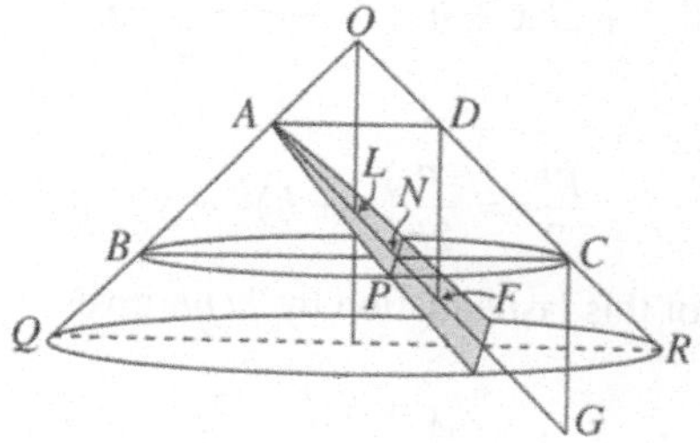


**1(c)** We have the following in the diagram for the solution to Exercise 1(a):

- $\measuredangle GNC \cong \measuredangle LAD$ because $\overline{AD}$ is parallel to $\overline{NC}$.
- $\overline{NC} \cong \overline{AD}$ since $\overline{AG}$ is parallel to $\overline{OR}$.
- $\measuredangle NCG \cong \measuredangle ADF$ because both are right.

Therefore, $\triangle NCG \cong \triangle ADF$ by SAS, so $\overline{NG} \cong \overline{AF}$. By parts (a) and (b) of this exercise, we conclude that

$$PN^2 = NC \cdot BN = NG \cdot NA = AF \cdot NA = 2AL \cdot NA.$$

**3** We will give two sample cones for the $y^2 = x$ equation.

- First use the cone from Exercise 1. The equation for that parabola is

$$PN^2 = 2AL \cdot NA,$$

where $PN = y$ and $NA = x$. This means that we want $AL = 1/2$. Since $\triangle AOL$ is isosceles with base $\overline{OL}$, to obtain this length we must cut the cone with a plane parallel to $\overline{OR}$ such that $AO = 1/2$.

- Next use the cone in the diagram for Theorem 7.3.4. We will assume that it is a right cone. In this case $PL = 1$. If the plane cuts the cone so that $PA = 1$ and $BP = 3$, then $BA = AC = 4$. Therefore, since

$$\frac{PL}{PA} = \frac{BC^2}{AC \cdot BA},$$

we have that $BC = 4$. This cone has a height of $2\sqrt{3}$, and its axial triangle is equilateral.

Of course, there are infinitely many cones that can be cut by a plane to yield the parabola $y^2 = x$.

**5** We start with

$$y^2 = x\left(\frac{p}{2h}x + p\right).$$

Next, we complete the square to find

$$y^2 = \frac{p}{2h}(x^2 + 2hx + h^2) - \frac{ph}{2} = \frac{p}{2h}(x+h)^2 - \frac{ph}{2}.$$

Therefore,

$$\frac{ph}{2} = \frac{p}{2h}(x+h)^2 - y^2.$$

Multiplying both sides of this last equation by $2/ph$ gives

$$1 = \frac{(x+h)^2}{h^2} - \frac{2y^2}{ph}.$$

**7** In Equation 7.13,

$$y^2 = x\left(\frac{p}{2h}x + p\right),$$

the square of the ordinate is $y^2$. The term on the right is the area of a rectangle with base $x$ and height $px/2h + p$. This height *overlaps* the parameter because

$$px/2h + p > p.$$

**9(b)** We will prove the result for an ellipse with diameter $\overline{PP'}$ and center $O$. Let $Q$ and $Q'$ be on the ellipse such that $\overline{QQ'}$ contains $O$. Let $V$ and $V'$ be on $\overline{PP'}$ such that $\overline{QV}$ and $\overline{Q'V'}$ are the ordinates paired, respectively, with abscissas $\overline{PV}$

and $\overline{PV'}$. We note that $\measuredangle QVO \cong \measuredangle Q'V'O$. By part (a) and using the fact that $\triangle OVQ \sim \triangle OV'Q'$,

$$\frac{PV \cdot P'V}{P'V' \cdot PV'} = \frac{QV^2}{(Q'V')^2} = \frac{OV^2}{(OV')^2}.$$

Therefore,

$$\frac{PV \cdot P'V}{OV^2} = \frac{P'V' \cdot PV'}{(OV')^2},$$

and then adding one to both sides yields

$$\frac{OP2 + PV \cdot P'V}{OP^2} = \frac{(OP')^2 + P'V' \cdot PV'}{(OP')^2}.$$

However, as on page 256 of the text,

$$PV \cdot P'V = (OV - OP)(OV + OP) = OV^2 - OP^2.$$

This gives

$$OP^2 + PV \cdot P'V = OV^2.$$

Similarly,

$$(OP')^2 + P'V' \cdot PV' = (OV')^2.$$

Hence,

$$\frac{OV^2}{OP^2} = \frac{(OV')^2}{(OP')^2}.$$

Therefore, $\overline{OV} \cong \overline{OV'}$ because $\overline{OP} \cong \overline{OP'}$. We finally conclude that $\triangle OVQ \cong \triangle OV'Q'$ by ASA, which gives $\overline{OQ} \cong \overline{OQ'}$.

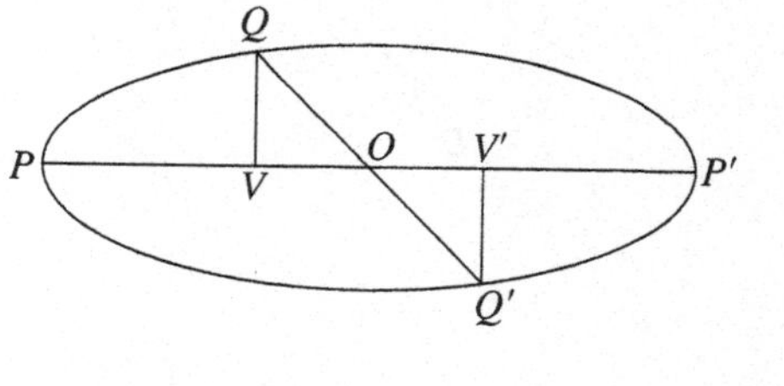


**11(b)** $p = 2b^2/a$

**13** This solution is due to Apollonius (*Conics* I.15). Let $\overline{PP'}$ be a diameter of the ellipse with center $O$. Let $\overline{AB}$ contain $O$ such that $\overline{OA}$ is an ordinate with abscissa $\overline{PO}$. Let $\overline{PL}$ be the parameter of the ordinates. Join $\overline{P'L}$ and let $R$, $R'$, and $E$ be on $\overline{P'L}$ such that $\overline{RV}$, $\overline{EO}$, and $\overline{R'V'}$ are parallel to $\overline{PL}$. Take any ordinate $\overline{QV}$ with $Q$ on the ellipse and $V$ on $\overline{PP'}$. Find $Q'$ on the ellipse so that $\overline{QQ'}$ is parallel to $\overline{PP'}$. Take $V'$ on $\overline{PP'}$ such that $\overline{Q'V'}$ is parallel to $\overline{QV}$. That is, $\overline{Q'V'}$ is another ordinate. We then have by Theorem 7.3.7 that

$$QV^2 = RV \cdot PV$$

and

$$(QV')^2 = R'V' \cdot PV'.$$

Since $\overline{QQ'}$ is parallel to $\overline{PP'}$,

$$\overline{QV} \cong \overline{QV'}.$$

Therefore,

$$RV \cdot PV = R'V' \cdot PV',$$

which implies that

$$\frac{PV}{PV'} = \frac{R'V'}{RV} = \frac{P'V'}{P'V},$$

where the second equality follows because $\triangle P'V'R' \sim \triangle P'VR$. Hence,

$$\frac{PV}{VV'} = \frac{P'V'}{VV'},$$

from which we conclude that $\overline{PV} \cong \overline{P'V'}$. Because $\overline{OP} \cong \overline{OP'}$, we have by segment subtraction that $\overline{OV} \cong \overline{OV'}$. In other words, $\overline{AB}$ bisects $\overline{VV'}$, and this implies that $\overline{AB}$ bisects $\overline{QQ'}$. Hence, $\overline{AB}$ is a diameter of the ellipse.

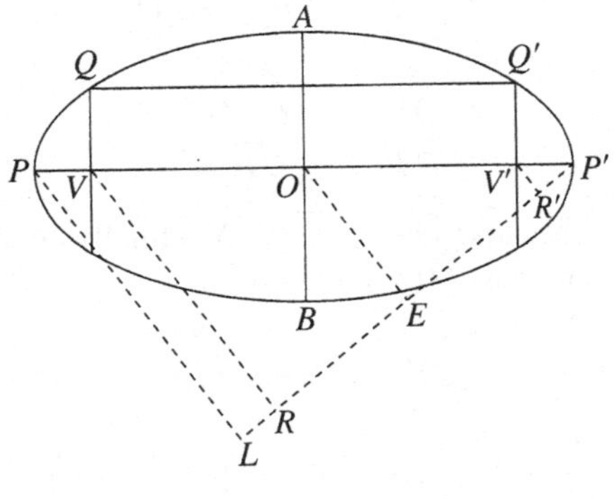


**15** $y = -\frac{1}{5}x$

## 7.4 THE ANALYTIC ART IN TWO VARIABLES

**1(a)** This is Fermat's solution: Start with the equation

$$ad = be.$$

Let $O$ be a point and find $A$ and $E$ making $\sphericalangle EOA$. Let $a = OA$ and $e = OE$. By assumption,

$$\frac{OA}{OE} = \frac{b}{d}.$$

Let $A'$ be any other point on the $\overline{OA}$ produced and choose $E'$ so that $\overline{A'E'}$ is parallel to $\overline{AE}$ and

$$\frac{OA'}{OE'} = \frac{b}{d}.$$

Therefore, $\triangle OAE \sim \triangle OA'E'$, which implies that $\sphericalangle EOA \cong \sphericalangle E'OA'$. Hence, all points $E'$ described in this manner lie on the same line.

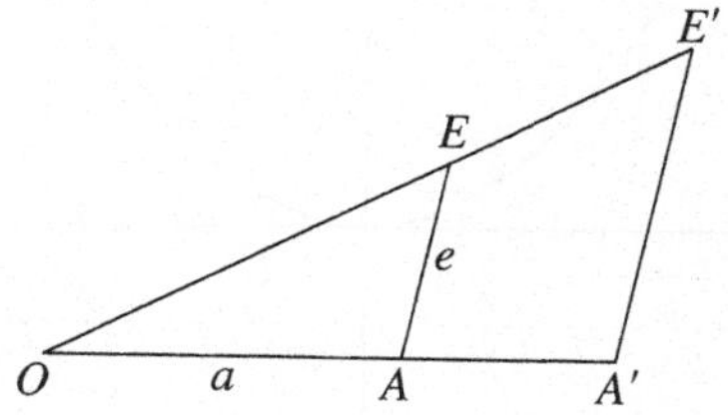


**1(c)** This is Fermat's solution: Let

$$b^2 - a^2 = e^2,$$

where $a$ and $e$ are unknowns. Fix $O$ and find $B$ such that $OB = b$. We will describe a circle with center $O$ and radius $OB$ using the given equation. Find $A$ between $O$ and $B$ and let $E$ be a point such that $\sphericalangle OAE$ is right and

$$b^2 - OA^2 = AE^2.$$

That is, $OA = a$ and $AE = e$. Therefore, $OE = b$ and we conclude that $E$ is on the first quadrant of a circle. The rest of the circle is described similarly.

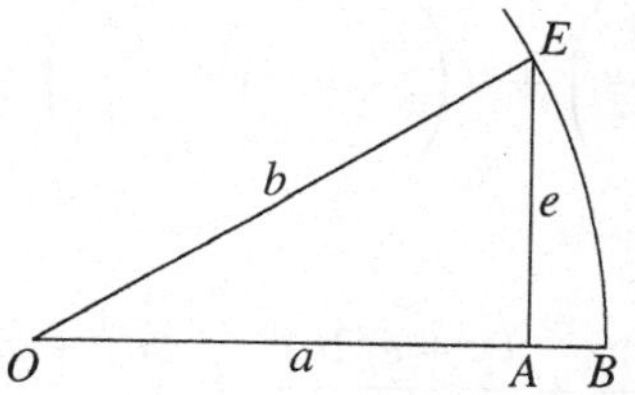


**2(b)** The graph of

$$y^2 = 4x + x^2$$

is a hyperbola with center $(-2, 0)$ and asymptote $y = x + 2$. Its graph with a 45° oblique ordinate is:

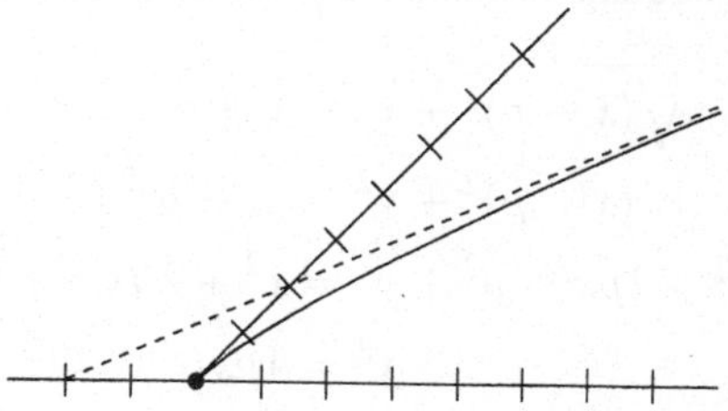

**3** For the conic section $y^2 = 4x + x^2$, the length of the parameter of the ordinates is 4 and the distance from the vertex to the focus is $2\sqrt{2} - 2$.

**4(b)** This describes a parabola that has the shape of the one described by $x^2 = 8y$.

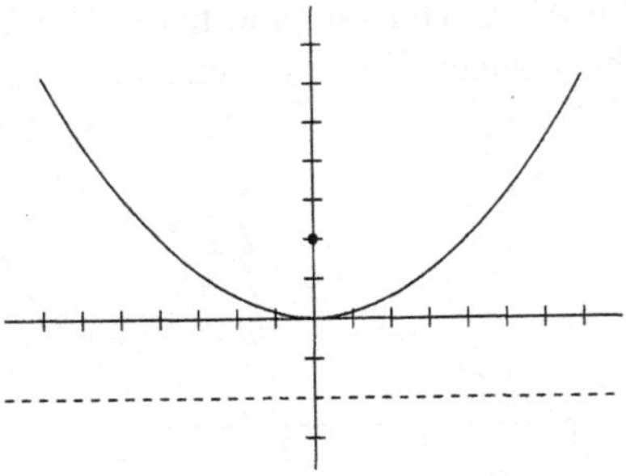

**5(a)** The graph is a point when it is a circle of radius 0 or a special case of an ellipse.

**5(c)** If the equation has at least two solutions and

$$c^2 - 4ab < 0,$$

the equation yields an ellipse. We may study this ellipse by first rotating it if necessary so that $c = 0$. The equation then becomes

$$ax^2 + by^2 + dx + ey + f = 0.$$

We may then complete the square on $x$ and $y$ separately to find further conditions:

$$a\left(x^2 + \frac{d}{a}x + \frac{d^2}{4a^2}\right) + b\left(y^2 + \frac{e}{b}y + \frac{e^2}{4b^2}\right) = -f + \frac{d^2}{4a} + \frac{e^2}{4b},$$

which yields

$$\frac{(x + d/2a)^2}{b} + \frac{(y + e/2b)^2}{a} = \frac{bd^2 + ae^2 - 4abf}{4a^2b^2}.$$

This will be an ellipse that is not a circle when $a \neq b$ and $bd^2 + ae^2 - 4abf > 0$.

**7** The vertex of this parabola is $(0, 0)$, its focus is at $(p, 0)$, and its directrix is $x = -p$. Let $(x, y)$ be any point on the parabola. We know that the distance from this point to the directrix equals its distance to the focus. That is,

$$\begin{aligned}\sqrt{(x-p)^2 + y^2} &= x + p,\\ (x-p)^2 + y^2 &= (x+p)^2,\\ x^2 - 2px + p^2 + y^2 &= x^2 + 2px + p^2,\\ y^2 &= 4px.\end{aligned}$$

**9** Let $2a$ be the length of the major axis, $2b$ the length of the minor axis, and $2c$ the distance between the foci. For this ellipse, $(0, 0)$ is the center, $(\pm c, 0)$ are the foci, $(\pm a, 0)$ are the vertices, and $(0, \pm b)$ are the endpoints of the minor axis. Let $k$ represent the constant distance. Observe that

$$k = 2\sqrt{b^2 + c^2}$$

and

$$k = (a - c) + (a + c) = 2a.$$

Therefore,

$$b^2 = a^2 - c^2.$$

Let $(x, y)$ be a point on the ellipse. We may now derive the equation:

$$\begin{aligned}
2a &= \sqrt{(x-c)^2 + y^2} + \sqrt{(x+c)^2 + y^2},\\
2a - \sqrt{(x+c)^2 + y^2} &= \sqrt{(x-c)^2 + y^2},\\
4a^2 - 4a\sqrt{(x+c)^2 + y^2} + (x+c)^2 + y^2 &= (x-c)^2 + y^2,\\
4a^2 - 4a\sqrt{(x+c)^2 + y^2} + 2cx &= -2cx,\\
4a\sqrt{(x+c)^2 + y^2} &= 4a^2 + 4cx,\\
a^2[(x+c)^2 + y^2] &= (a^2 + cx)^2,\\
a^2(x^2 + 2cx + c^2 + y^2) &= a^4 + 2a^2cx + c^2x^2,\\
(a^2 - c^2)x^2 + a^2y^2 &= a^2(a^2 - c^2),\\
b^2x^2 + a^2y^2 &= a^2b^2.
\end{aligned}$$

The last equation is equivalent to

$$\frac{x^2}{a^2} + \frac{y^2}{b^2} = 1.$$

**11** The focus of this parabola is $(2, 3)$, $P$ is $(2, 7)$, and $P'$ is $(2, -1)$. For the parabola,

$$\frac{dy}{dx} = \frac{8}{2y - 6},$$

so the tangent line at $P$ is $y = x + 5$ and the tangent line at $P'$ is $y = -x + 1$. These lines meet $y = 3$ at $(-2, 3)$.

**13** It suffices to show that the property holds at $P_1$. Let the parabola have equation $x^2 = 4py$ so that the vertex is $(0, 0)$ and the focus is $F(0, p)$. Let $(x_0, y_0)$ be the coordinates of $P_1$. This means that

$$y_0 = \frac{x_0^2}{4p}.$$

Notice that

$$\frac{dy}{dx} = \frac{x}{2p},$$

so that the equation of the tangent line to the parabola at $P_1$ is

$$y = \frac{x_0}{2p}x - \frac{x_0^2}{4p}.$$

Let $S_1$ be the $y$-intercept of the tangent line, so the coordinates of $S_1$ are $(0, -y_0)$. Therefore,

$$S_1 F = y_0 + p.$$

Furthermore,

$$\begin{aligned} P_1 F &= \sqrt{x_0^2 + (y_0 - p)^2} \\ &= \sqrt{x_0^2 + y_0^2 - 2py_0 + p^2} \\ &= \sqrt{4py_0 + y_0^2 - 2py_0 + p^2} \\ &= \sqrt{y_0^2 + 2py_0 + p^2} \\ &= y_0 + p. \end{aligned}$$

Therefore, $\triangle FP_1S_1$ is an isosceles triangle. We then have $\measuredangle FP_1S_1 \cong \measuredangle FS_1P_1$. Since $\overline{Q_1P_1}$ is parallel to the $y$-axis, $\measuredangle R_1P_1Q_1 \cong \measuredangle FS_1P_1$. Hence, $\measuredangle R_1P_1Q_1 \cong \measuredangle FP_1S_1$.

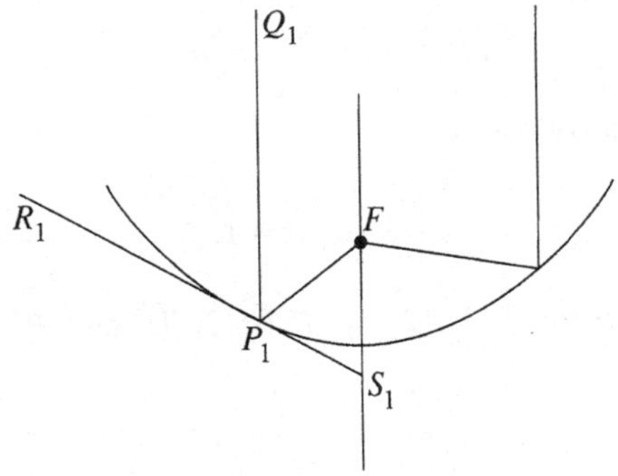

# CHAPTER 8

# RENÉ DESCARTES

## 8.1 COMPASSES

**1** To solve $x^2 = 2$, we need only two of the triangles, so slide the rightmost triangle to the side. Let $AE = 2$ and slide the remaining movable triangle so that $CG = 1$ and $\overline{AK}$, $\overline{MF}$, and $\overline{M'G}$ meet at $B$. Then, as in the text,

$$\frac{2}{BF} = \frac{BF}{1},$$

which implies that $BF^2 = 2$. Hence, $BF$ is the solution to $x^2 = 2$.

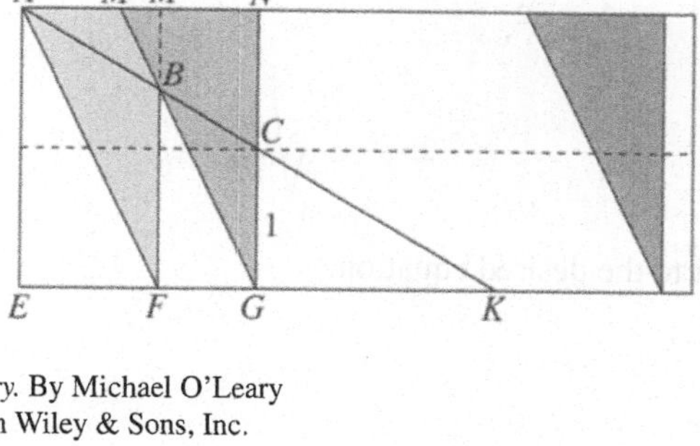

**3(a)** A negative root.

**3(b)** It has 2, 3, and 4 as roots, and 5 as a false root.

**4(a)** By Descartes Rule of Signs, this equation has either 2 or 0 real solutions, and 1 negative (false) solution. Since $y = x^3$ and $y = x - 4$ can intersect at most once, we conclude that it has only the false solution, which cannot be found by Descartes' mesolabe.

**5(a)** We must first set up a coordinate system. Let $Y$ be the origin and $\overline{YZ}$ be the $x$-axis. We will use ordinates that meet the $x$-axis in right angles. Let $r = YA$. By construction, $\overline{YB} \cong \overline{YA}$. As the mesolabe opens with pivot at $Y$, $B$ traces out a circle. Therefore, the equation for the curve through $A$ and $B$ is $x^2 + y^2 = r^2$.

**5(c)** We use the same coordinate system as before. Let $YA = YB = r$, $YE = x$, $EF = y$, and $YF = z$. Therefore, by the construction of the mesolabe, $\Delta YEF$ is similar to $\Delta YDE$, so

$$\frac{z}{x} = \frac{x}{YD},$$

and then

$$YD = \frac{x^2}{z}.$$

Also,

$$\frac{x}{YD} = \frac{YD}{YC}$$

because $\Delta YDE \sim \Delta YCD$, so

$$YC = \frac{YD^2}{x} = \frac{x^3}{z^2}.$$

Since $YD/YC = YC/r$,

$$\frac{rx^2}{z} = \frac{x^6}{z^4}.$$

Therefore,

$$z = \sqrt[3]{\frac{x^4}{r}},$$

and then because $z^2 = x^2 + y^2$,

$$x^2 + y^2 = \sqrt[3]{\frac{x^8}{r^2}},$$

which is equivalent to the desired equation.

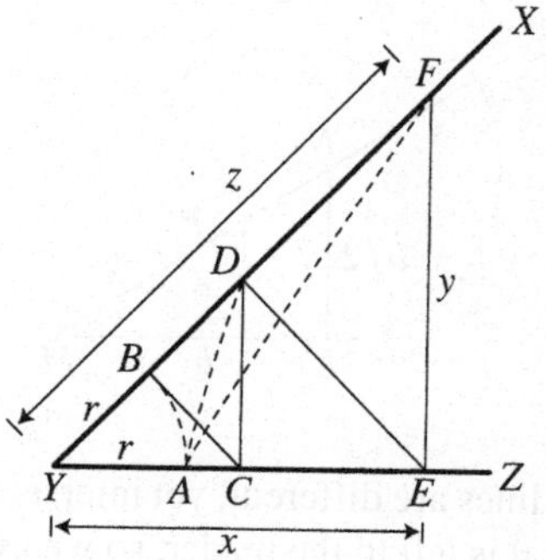


## 8.2 METHOD

The are no exercises for this section.

## 8.3 ANALYTIC GEOMETRY

**1** The top diagrams show how to construct $5 + 3$ and $5 - 3$. In the triangle on the left, $AE = 15$, and in the triangle on the right, $AD = 5/3$.

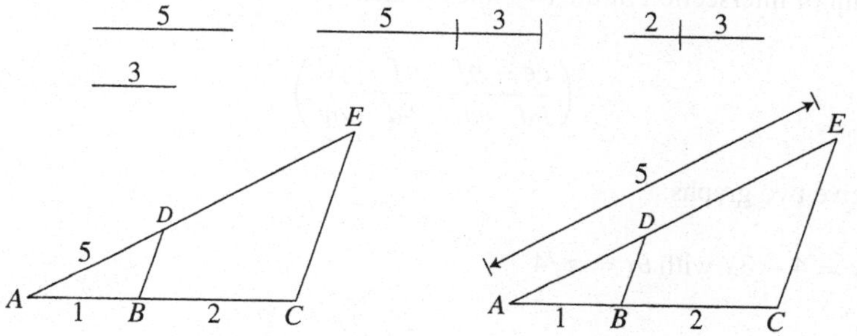


**3(b)** Construct right $\triangle NLM$ with $\measuredangle NLM$ right, $LM = b$, and $NL = a/2$. Let $x = PM$. Therefore, by the Tangent-Secant Theorem (5.5.5),

$$b^2 = x(a + x) = ax + x^2,$$

which is equivalent to

$$x^2 = b^2 - ax.$$

Hence, $PM$ is the solution to the equation, and this is

$$x = NM - NP = -\frac{a}{2} + \sqrt{\frac{a^2}{4} + b^2}.$$

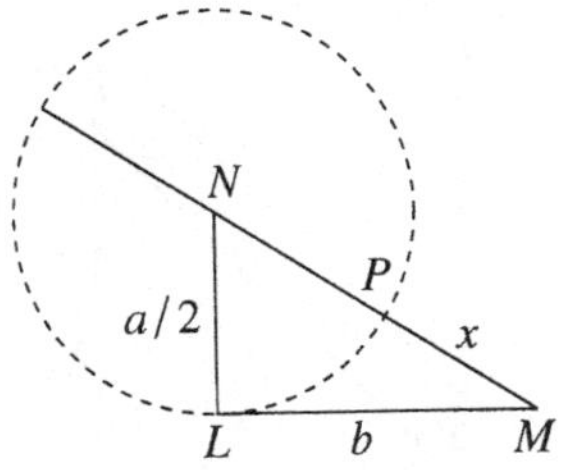


**5(a)** Suppose that the two lines are different, yet intersect. The case when one of the lines is horizontal or vertical is left to the reader, so we assume that $a$, $b$, $d$, and $e$ are nonzero. Let

$$ax + by + c = 0$$

and

$$dx + ey + f = 0$$

be the equations for the lines. By hypothesis, their slopes are different, so

$$bd - ae \neq 0.$$

The point of intersection of the two lines is then

$$\left(\frac{ce - bf}{bd - ae}, \frac{af - cd}{bd - ae}\right).$$

**8** We give two graphs:

- $y = 4 - 3x$ with $\theta_1 = \pi/4$

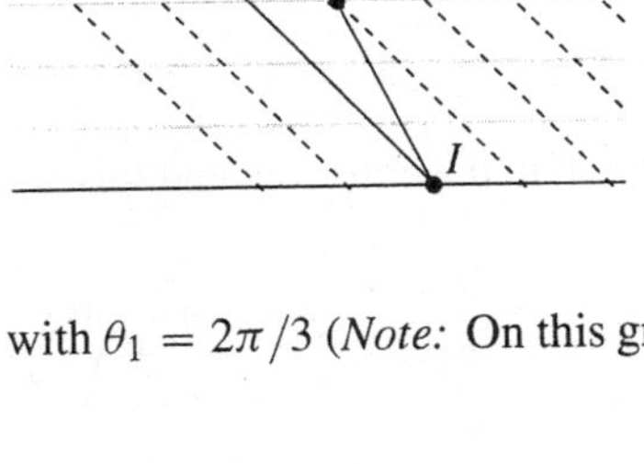


- $y = 1 + x + \sqrt{1 - x^2}$ with $\theta_1 = 2\pi/3$ (*Note:* On this graph the intervals have length .25.)

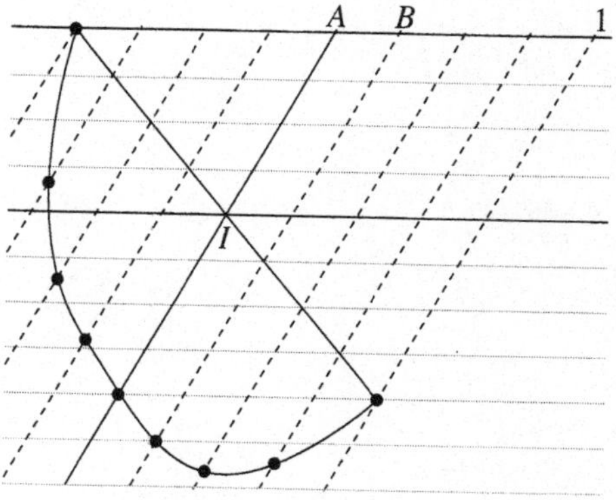


**9(a)** Modify Descartes' proof for this condition and use the given diagram.

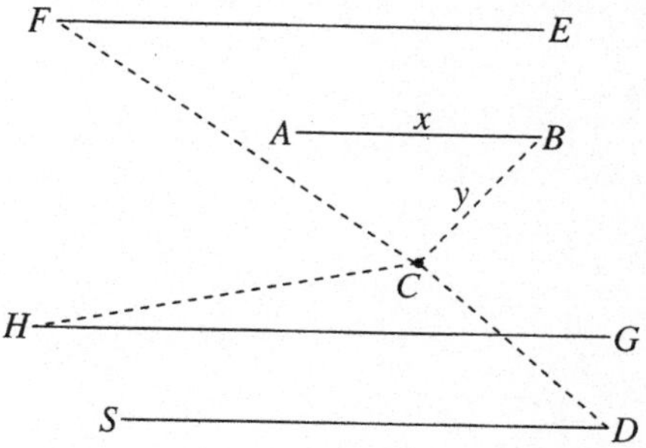


**11** We note the $b = k + x$, $c = k - x$, and $d = 2k - x$. Therefore,

$$kxy = bcd = (k + x)(k - x)(2k - x).$$

# CHAPTER 9

# GÉRARD DESARGUES

## 9.1 PROJECTIONS

The are no exercises for this section.

## 9.2 POINTS AT INFINITY

**1** A parallelogram is a quadrilateral with parallel opposite sides. In $ABCD$, $\overline{AB}$ is parallel to $\overline{DC}$ because produced they intersect at ideal point $\Omega$, and we also see that $\overline{BC}$ and $\overline{AD}$ are parallel because they meet at $\Omega'$. A cylinder in projective space is a frustum of a cone where the cone's vertex is an ideal point.

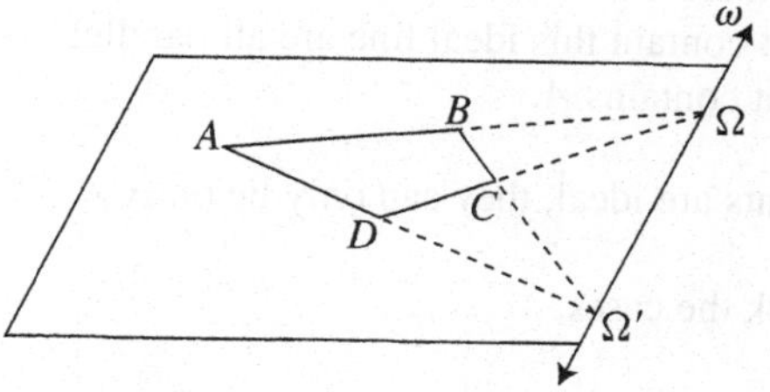

**3** Let $O$ and $O'$ be distinct points in a projective plane. An ideal point is added to each line through $O$. Let $\Omega$ be the ideal point on $l$ so defined. By Postulate 9.2.5, there exists $l'$ on $O'$ parallel to $l$. This means that $l$ and $l'$ intersect at $\Omega$. If we choose the ideal points for the lines through $O'$ in such a manner that the ideal point for $l'$ is different from $\Omega$, there would be two distinct lines common to a pair of distinct points contradicting Theorem 9.2.7. Therefore, the same set of ideal points are defined using both $O$ and $O'$, and thus the same ideal line is generated by both points.

**5** Let $l$ and $l'$ be two different lines in a projective plane.

- By Theorem 9.2.7, we know that the two lines cannot have more than one point in common.

- Since $l$ and $l'$ either intersect in an ordinary point (if they are not parallel) or in an ideal point (if they are parallel), they intersect in at least one point.

Therefore, $l$ and $l'$ intersect in exactly one point.

**6(a)** If either both points are ordinary or one is ordinary and the other ideal, there is a projective plane that contains both points. This means that we can apply Theorem 9.2.7. Suppose that both points are ideal. Call them $\Omega_l$ and $\Omega_{l'}$, where these points are found on distinct lines $l$ and $l'$ through some point $O$. Let $\Lambda$ be the unique projective plane that contains both lines and let $\omega$ be the ideal line of the plane $\Lambda$. By Postulate 9.2.10.2, $\omega$ contains both $\Omega_l$ and $\Omega_{l'}$. This line is unique since in the construction of projective space, exactly one ideal line is added per projective plane.

**6(d)** Choose three noncollinear points. We have numerous cases to check.

- If the three points are ordinary, they determine a plane uniquely. Moreover, they are not on the ideal plane by Postulate 9.2.10.5.

- Suppose that exactly two of the points are ordinary. Let $A$ and $B$ be the two ordinary points and $\Omega_l$ be the ideal point, where $l$ is the line through $O$ containing $\Omega_l$. Join $\overline{AB}$. Because the points are not collinear, $\overline{AB}$ and $l$ intersect in an ordinary point forming an angle, and this angle describes a plane uniquely.

- Now assume that exactly one of the three points is ordinary. Let $A$ be the ordinary point and $\Omega_l$ and $\Omega_m$ be the ideal points, with $l$ and $m$ on $O$. By Exercise 6(a), there exists exactly one line $\omega$ incident with both $\Omega_l$ and $\Omega_m$. The planes that contain this ideal line are all parallel, so the one that we want is the plane that contains $A$.

- If all three points are ideal, they can only lie on $\Lambda_\infty$.

**6(h)** Again, we check the cases.

- If both the plane and line are ordinary, we know from Euclidean geometry that the line either intersects the plane in exactly one ordinary point or the line and plane are parallel. If the latter is the case, they meet in exactly one ideal point.
- Suppose that the plane $\Lambda$ is ordinary but the line $\omega$ is ideal.

**7** Let $\Omega$ be on $\Lambda$ and $l$. Suppose that the point $A$ is on $l$ and on $\Lambda$. By Exercise 6(h), $l$ must be on $\Lambda$.

## 9.3 THEOREMS OF DESARGUES AND MENELAUS

**1** We use Theorem 9.3.1 to prove this. Suppose that the corresponding sides of $\triangle ABC$ and $\triangle A'B'C'$ intersect in collinear points. We need to show that the join of corresponding vertices are concurrent. Let $O$ (not pictured) be the intersection of $\overline{AA'}$ and $\overline{CC'}$. We need to show that $O$ is on $\overline{BB'}$. Consider $\triangle QAA'$ and $\triangle RCC'$. Observe that $\overline{A'C'}$, $\overline{AC}$, and $\overline{QR}$ intersect at $P$. Thus, by Theorem 9.3.1, the corresponding sides of these triangles meet in collinear points. However, $\overline{QA}$ intersects $\overline{RC}$ at $B$, $\overline{QA'}$ intersects $\overline{RC'}$ at $B'$, and $\overline{AA'}$ intersects $\overline{CC'}$ at $O$. Hence, $O$ is on $\overline{BB'}$.

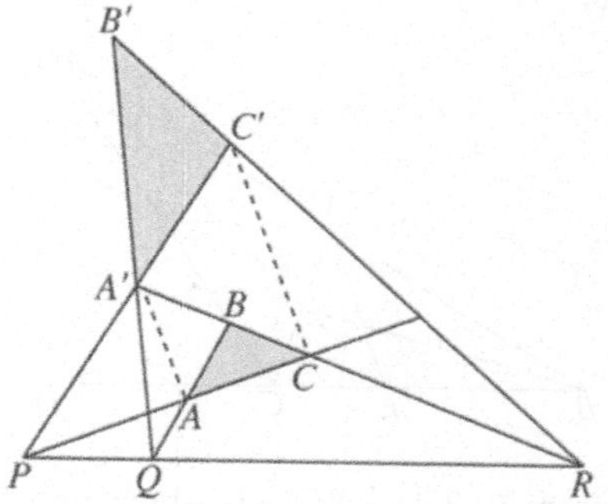


**3** We are given $\triangle ABC$. Let $D$ be on the line through $B$ and $C$, $E$ on the line through $A$ and $C$, and $F$ on the line through $A$ and $B$. Suppose that all of the points $D, E, F$ are not on the triangle. Find $G$, $I$, and $H$ on $\overline{FD}$ such that $\overline{BG}$, $\overline{CI}$, and $\overline{AH}$ are all perpendicular to $\overline{FD}$. We can now proceed as in the proof given for the first case of Theorem 9.3.3.

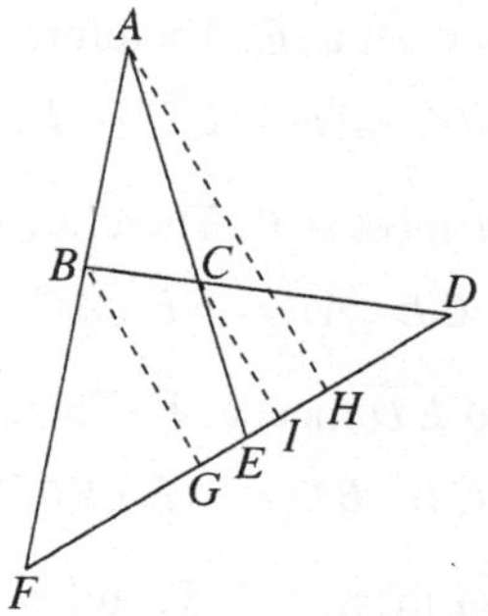

**5** Let $\triangle ABC$ be given. Extend $\overline{BC}$ to $D$ to form external $\measuredangle ACD$ and find $G$ so that $\overline{CG}$ bisects $\measuredangle ACD$. Also, find $E$ on $\overline{BC}$ and $F$ on $\overline{AC}$ so that $\overline{AE}$ bisects $\measuredangle BAC$ and $\overline{BF}$ bisects $\measuredangle ABC$. If we can show that

$$\sin \measuredangle ACG \sin \measuredangle BAE \sin \measuredangle CBF = \sin \measuredangle GCB \sin \measuredangle EAC \sin \measuredangle FBA,$$

we may then apply Exercise 4 to obtain the result. First note that since

$$\measuredangle CBF \cong \measuredangle FBA$$

and

$$\measuredangle BAE \cong \measuredangle EAC,$$

we have

$$\sin \measuredangle CBF = \sin \measuredangle FBA \text{ and } \sin \measuredangle BAE = \sin \measuredangle EAC.$$

Furthermore, letting $\alpha = \text{m}\measuredangle ACG$, $\beta = \text{m}\measuredangle GCB$, and $\delta = \text{m}\measuredangle GCD$,

$$\sin\alpha = \sin(\pi - \alpha) = \sin(\pi - \delta) = \sin\beta.$$

Hence, the result follows.

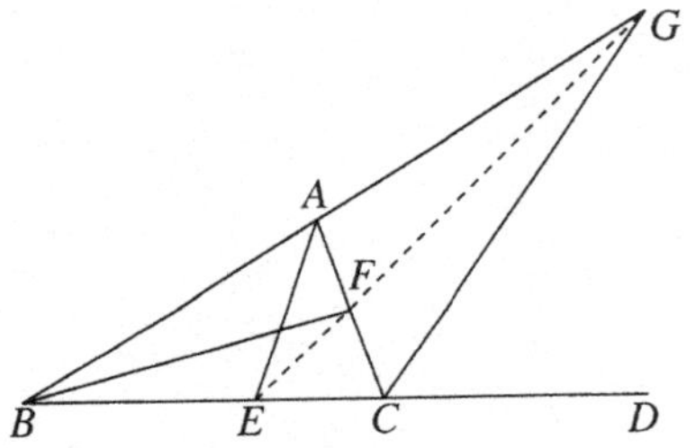


**7** Assume Ceva's Theorem (9.3.4). We will show one direction of Menelaus's Theorem (9.3.3). In $\triangle ABC$, let $F$ be on $\overline{AB}$, $E$ on $\overline{AC}$, and $D$ on $\overline{BC}$ beyond $C$. Suppose that these three points are collinear. We need some auxiliary lines. Join $\overline{CF}$ and $\overline{AD}$ and let $H$ be their intersection. Join $\overline{BE}$ and let $G$ be its intersection with $\overline{CF}$ and $I$ its intersection with $\overline{AD}$.. We now apply Ceva's Theorem and make the following observations:

- In $\triangle ABD$, $\overline{AC}$, $\overline{BI}$, $\overline{FD}$ meet at $E$. Therefore,
$$BC \cdot ID \cdot AF = CD \cdot AI \cdot BF.$$
- In $\triangle ADE$, $\overline{AF}$, $\overline{CD}$, $\overline{EI}$ meet at $B$. Therefore,
$$FD \cdot CE \cdot AI = FE \cdot AC \cdot ID.$$
- In $\triangle BCE$, $\overline{AB}$, $\overline{CG}$, and $\overline{ED}$ meet at $F$. Therefore,
$$AC \cdot GE \cdot BD = AE \cdot BG \cdot CD.$$
- In $\triangle BEF$, $\overline{BD}$, $\overline{AE}$, $\overline{FG}$ meet at $C$. Therefore,

$$ED \cdot AF \cdot BG = FD \cdot AB \cdot GE.$$

- In $\triangle CAF$, $\overline{BC}$, $\overline{AH}$, $\overline{FE}$ meet at $D$. Therefore,

$$AB \cdot FH \cdot CE = BF \cdot CH \cdot AE.$$

- In $\triangle CFD$, $\overline{CE}$, $\overline{BF}$, $\overline{DH}$ meet at $A$. Therefore,

$$FE \cdot BD \cdot CH = ED \cdot BC \cdot FH.$$

Therefore,

$$BF^2 \cdot AE^2 \cdot CD^2 = AF^2 \cdot CE^2 \cdot BD^2,$$

which implies that

$$BF \cdot AE \cdot CD = AF \cdot CE \cdot BD.$$

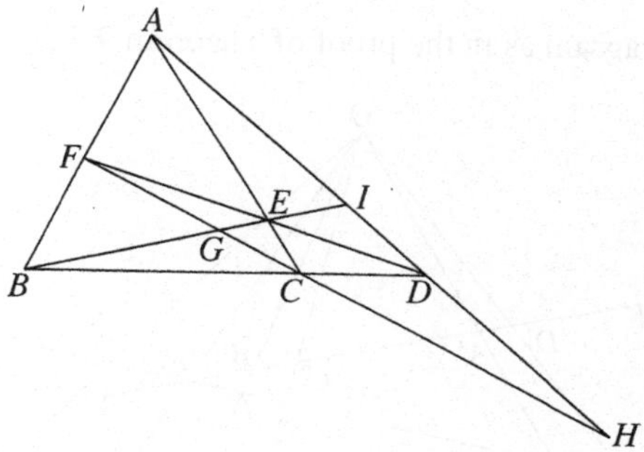


**9(a)** In $\triangle ABC$, let $D$ be the midpoint of $\overline{BC}$, $E$ the midpoint of $\overline{AC}$, and $F$ the midpoint of $\overline{AB}$. Therefore,

$$AF = BF, BD = CD, \text{ and } CE = AE,$$

and this implies that

$$AF \cdot BD \cdot CE = BF \cdot CD \cdot AE.$$

By Ceva's Theorem (9.3.4), the medians are concurrent.

**11** Since Pappus's Theorem is a generalization of Pascal's Theorem, modify appropriately the proof for Theorem 9.3.5.

## 9.4 INVOLUTIONS

**1(a)** Let $A$, $B$; $C$, $D$; and $E$, $F$ be three pairs of collinear points that form a tree with stump $O$. We will check the case when $O$ separates the pairs so that $A$, $C$, and $E$ are on one side of $O$ and $B$, $D$, and $F$ are on the other side. So,

$$\frac{DE}{CF} = \frac{OE + OD}{OC + OF} = \frac{OD}{OF}.$$

The second equality holds because

$$OE \cdot OF + OD \cdot OF = OC \cdot OD + OF \cdot OD.$$

Similar arguments show the other equations.

**3** Assume that $(AB)(CD)(EF)$. Then $(EF)(CD)(AB)$ follows easily. To show $(AB)(EF)(CD)$, first note that $\mathsf{T}(AB)(CD)(EF)$ follows by Exercise 2. By Exercise 1 and the symmetry of the definition of a tree,

$$\frac{OE}{OF} = \frac{EA \cdot EB}{FA \cdot FB} = \frac{EC \cdot ED}{FC \cdot FD}.$$

Therefore, $(AB)(EF)(CD)$.

**5** We use the same diagram as in the proof of Theorem 9.4.2:

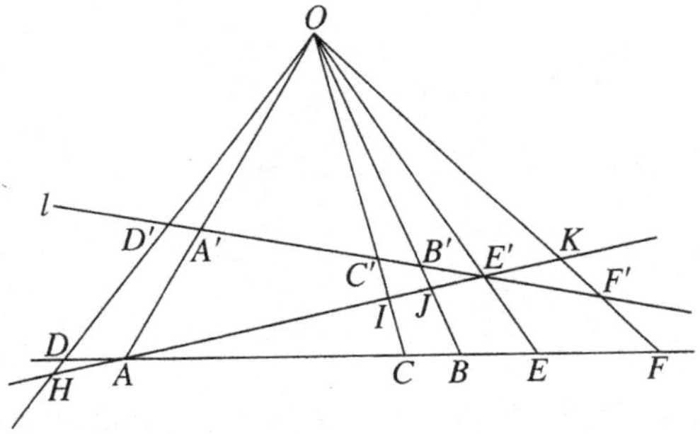


In $\triangle E'A'A$ with transversal $\overline{D'H}$, we have by Menelaus's Theorem that

$$E'D' \cdot AH \cdot A'O = A'D' \cdot E'H \cdot AO,$$

and in the same triangle with transversal $\overline{KF'}$, we have that

$$E'F' \cdot AK \cdot A'O = A'F' \cdot E'K \cdot AO.$$

Therefore,

$$\frac{A'D'}{E'D'} = \frac{AH \cdot A'O}{E'H \cdot AO}$$

and

$$\frac{A'F'}{E'F'} = \frac{AK \cdot A'O}{E'K \cdot AO}.$$

Now using $\triangle AEE'$ with transversal $\overline{HD}$, we have

$$AD \cdot E'H \cdot EO = ED \cdot AH \cdot E'O,$$

and with transversal $\overline{FK}$ we obtain

$$AF \cdot E'K \cdot EO = EF \cdot AK \cdot E'O.$$

Hence,

$$\frac{AH}{E'H} = \frac{AD \cdot EO}{ED \cdot E'O}$$

and

$$\frac{AK}{E'K} = \frac{AF \cdot EO}{EF \cdot E'O}.$$

Wrapping this up we find that

$$\frac{A'D' \cdot A'F'}{E'D' \cdot E'F'} = \frac{A'O^2 \cdot AH \cdot AK}{AO^2 \cdot E'H \cdot E'K} = \frac{A'O^2 \cdot EO^2 \cdot AD \cdot AF}{AO^2 \cdot E'O^2 \cdot ED \cdot EF}.$$

**7** Of course, a random set of six collinear points are not guaranteed to form a quadrangular set. Instead, given five collinear points, we can find a sixth point so that the set of six points forms a quadrangular set. Let $D$, $A$, $C$, $B$, and $E$ be collinear as in the diagram. To construct the complete quadrangle, follow these steps:

- Draw $\overline{QB}$ and $\overline{TE}$ intersecting at $N$.
- Draw $\overline{CT}$ intersecting $\overline{QB}$ at $R$.
- Draw $\overline{AR}$ extending it to meet $\overline{TE}$ at $S$.
- Draw $\overline{DT}$ intersecting $\overline{QB}$ at $Q$. (Since $Q$ is simply an endpoint, it can be moved if necessary.) Let $M$ be the intersection of $\overline{AS}$ and $\overline{DT}$.
- Join $\overline{QS}$ and extend it to meet $\overline{DE}$ at some $F$. Note that $F$ might be the ideal point of $\overline{DE}$.

The diagonal triangle of this complete quadrangle is $\triangle PMN$.

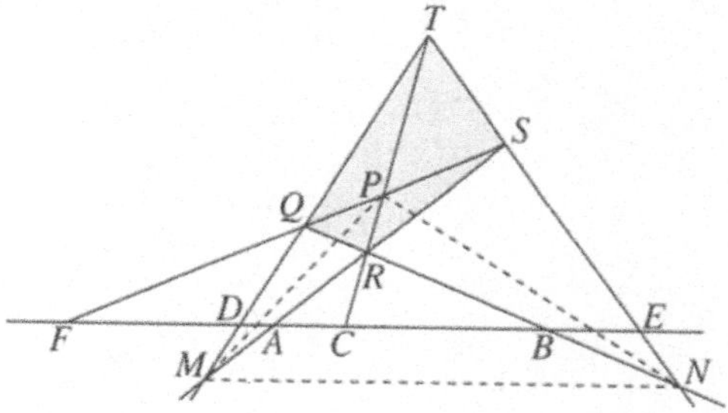


**9(a)** In the diagram, points $D$, $E$, and $F$ are the shared triangle triple of points.

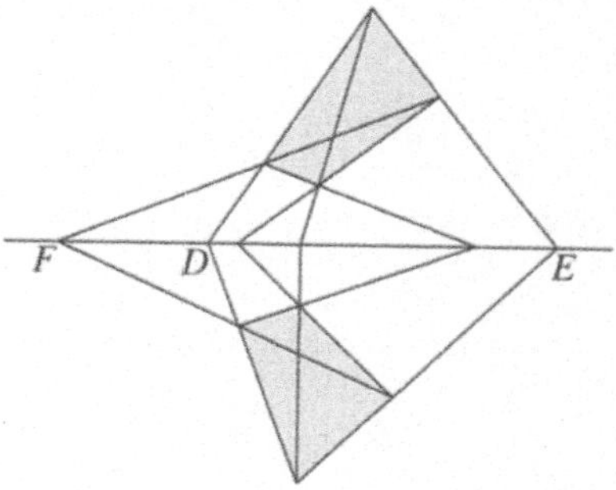

**11** How to find $F$ is explained in the solution to Exercise 7. The point $F$ is unique. The proof of this is a modification of the proof of Theorem 14.2.9.

**13** We first note that the construction is not unique. It depends upon the choice of the first vertex. Let $\triangle ABC$ be the diagonal triangle and let $T$ be chosen as one of the vertices of the complete quadrangle. Let $A'$ be the intersection of $\overline{TA}$ and $\overline{BC}$, $B'$ be the intersection of $\overline{TB}$ and $\overline{AC}$, and $C'$ be the intersection of $\overline{TC}$ and $\overline{BA}$. (Notice that the lines joining corresponding vertices of $\triangle ABC$ and $\triangle A'B'C'$ intersect at $T$.) Let $D$ be the intersection of $\overline{A'B'}$ and $\overline{BC'}$ and $E$ be the intersection of $\overline{A'C'}$ and $\overline{B'C}$. Finally, let $Q$ be the intersection of $\overline{CD}$ and $\overline{TB}$, and let $S$ be the intersection of $\overline{BE}$ and $\overline{TC}$. Then $\overline{BS}$, $\overline{QC}$, and $\overline{TA}$ intersect at $R$. This results in complete quadrilateral $QRST$. Try this construction on various diagonal triangles and vertices.

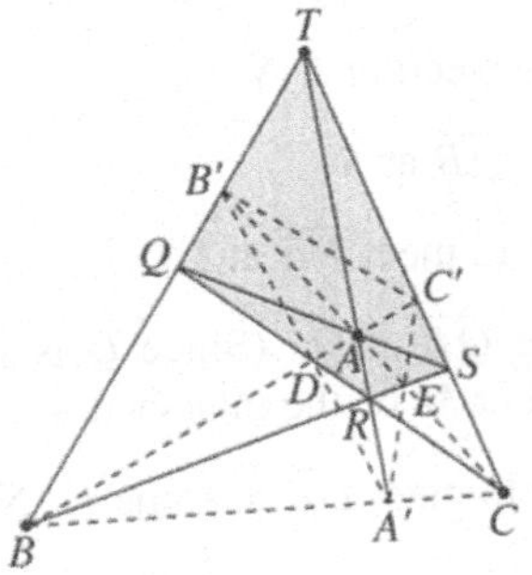

# CHAPTER 10

# GIOVANNI SACCHERI

## 10.1 THE QUESTION OF PARALLELS

**1(a)** Let $\overline{AC}$ and $\overline{BD}$ be perpendicular to $\overline{AB}$. Suppose that $\overline{AC}$ and $\overline{BD}$ are not equidistant. This means that we may assume that $\overline{CD}$ is perpendicular to $\overline{BD}$ but $AB \neq CD$. First suppose that $CD < AB$. By the replacement to the Parallel Postulate, there exists $\overline{AE}$ such that $\overline{AE}$ is equidistant to $\overline{BD}$, and we can then extend $\overline{CD}$ to intersect $\overline{AE}$ at $F$. Observe the following:

- $\overline{AB} \cong \overline{FD}$.
- $\sphericalangle BAF$ must be obtuse.

Let $M$ be the midpoint of $\overline{BD}$ and $N$ the midpoint of $\overline{AF}$. As in the proof of Theorem 10.2.2, we conclude that $\sphericalangle MNF$ is a right angle. However, since $\overline{AF}$ is equidistant to $\overline{BD}$, we have that $\overline{NM} \cong \overline{FD}$. This is impossible because the method found in the proof of Theorem 10.2.1 demonstrates the following:

- $\sphericalangle BAF \cong \sphericalangle AFD$ (showing that $\sphericalangle AFD$ is obtuse).
- $\sphericalangle MNF \cong \sphericalangle AFD$ (showing that $\sphericalangle AFD$ is right).

The case when $CD > AB$ is proven similarly.

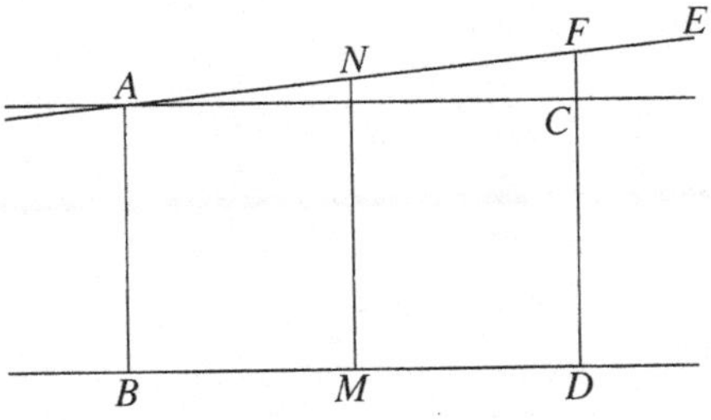


**1(b)** We will show sufficiency. Let $\overline{GC}$ and $\overline{DE}$ be equidistant. Let transversal $\overline{AF}$ intersect $\overline{GC}$ and $\overline{DE}$, respectively, at $B$ and $D$. If $\sphericalangle ABC$ is right, $\sphericalangle BDH$ is also right since the lines are equidistant, and the result is proven. So, suppose that $\sphericalangle ABC$ is acute. (The obtuse case is similarly proven.) Find $H$ on $\overline{DE}$ such that $\overline{BH}$ is perpendicular to $\overline{DE}$, and let $\overline{DG}$ be perpendicular to $\overline{GC}$. Because the lines are equidistant, $\overline{DG} \cong \overline{BH}$, and by Exercise 1(a), we conclude that $\overline{GB} \cong \overline{DH}$. Therefore,

$$\triangle GBD \cong \triangle HDB$$

by SAS or SSS. Hence,

$$\sphericalangle BDH \cong \sphericalangle GBD \cong \sphericalangle ABC.$$

The other corresponding angles can also be shown to be congruent.

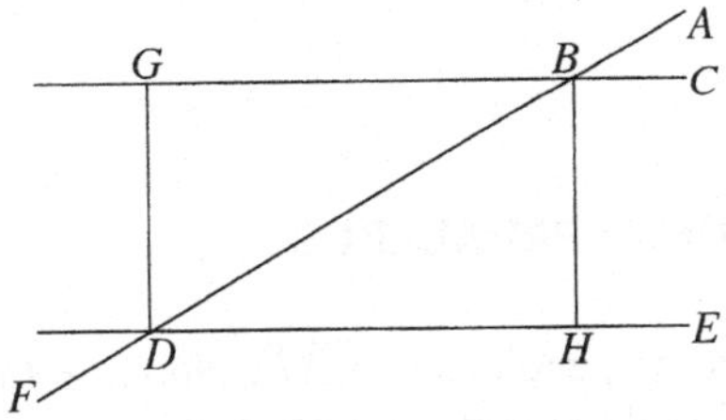


**3** Let transversal $\overline{AC}$ cut the parallel lines $\overline{AB}$ and $\overline{DC}$. Suppose that the alternate interior $\sphericalangle BAC$ and $\sphericalangle ACD$ are not congruent. We may then find a point $E$ such that $\sphericalangle EAC \cong \sphericalangle ACD$. But then $\overline{AE}$ is parallel to $\overline{DC}$ by Theorem 5.4.1, so we have two different lines through $A$ parallel to $\overline{DC}$, contradicting Playfair's Postulate (10.1.1).

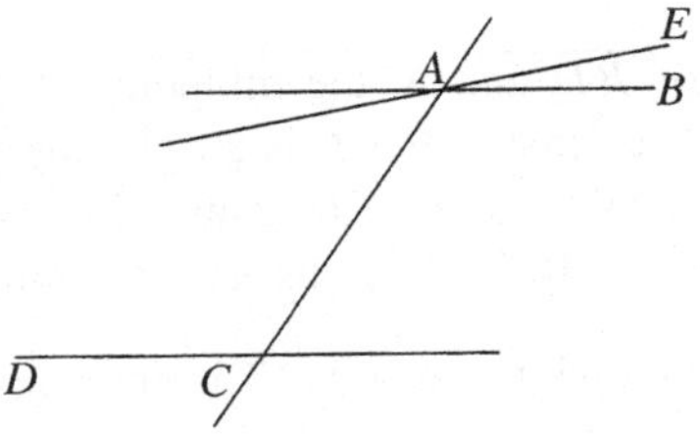

**5(a)** If $F = G$, then in $\triangle AGC$ we have $AC = CG + AG$, which is impossible (Theorem 5.3.11). The same goes for the other pairs of points.

**7** Assume that parallel lines are not equidistant. Let $\overline{AB}$ and $\overline{AD}$ be perpendicular and then construct $\overline{DC}$ perpendicular to $\overline{AD}$. Since $\overline{AB}$ and $\overline{DC}$ are parallel (Theorem 5.4.1), they are not equidistant, so we can find $E$ on $\overline{AB}$ and $F$ on $\overline{DC}$ such that $\overline{FE}$ is perpendicular to $\overline{AB}$ but $AD \neq FE$. Assume that $AD > FE$ and find $G$ on $\overline{AD}$ such that $\overline{AG} \cong \overline{EF}$. One can show that $\sphericalangle AGF \cong \sphericalangle EFG$. However, in $\triangle DFG$ we see that $\sphericalangle DGF$ must be acute because $\sphericalangle GDF$ is right (Theorem 5.3.12). Hence, $\sphericalangle AGF$ is obtuse and then so is $\sphericalangle GFE$. We conclude that $\sphericalangle DFE$ is obtuse and we have constructed a quadrilateral that has exactly three right angles, and thus it is not a rectangle. This quadrilateral is also not a rectangle when $AD < FE$.

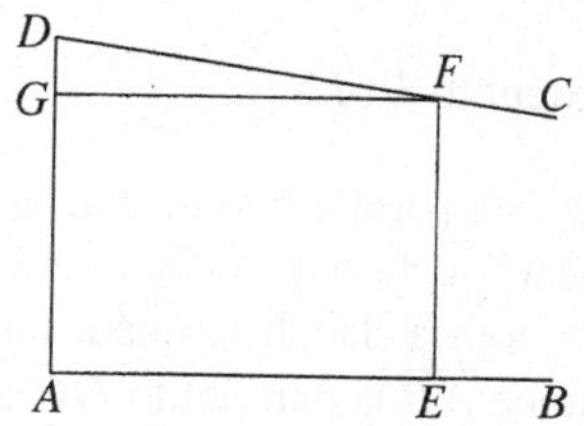


## 10.2 THE THREE HYPOTHESES

**1(a)** $AD = 6$, but the length of the summit $DC$ cannot be determined.

**1(c)** $m\sphericalangle ADC = \pi/4$

**1(e)** Cannot be determined.

**3(b)** In Saccheri quadrilateral $ABCD$, let $M$ be the midpoint of $\overline{AB}$ and $N$ the midpoint of $\overline{DC}$. Then, $\overline{MN}$ is perpendicular to both $\overline{AB}$ and $\overline{DC}$ (Theorem 10.2.2). Since the internal angles on the same side of $\overline{MN}$ are supplementary, $\overline{AB}$ does not intersect $\overline{DC}$ by Theorem 5.4.1. (*Remember:* Theorem 5.4.1 does not require the Parallel Postulate in its proof.)

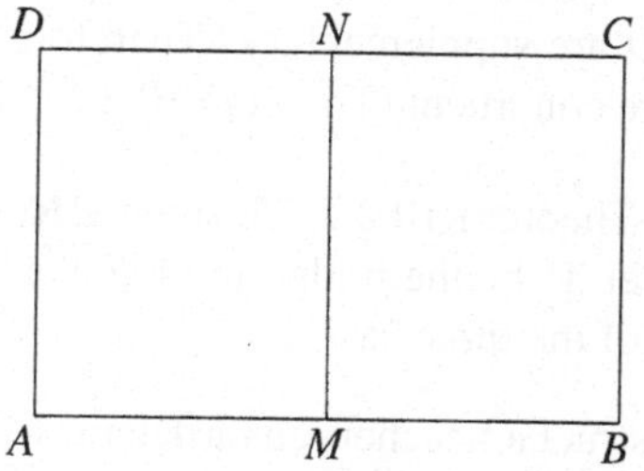


**3(c)** Continuing with the notation from the previous solution, in Saccheri quadrilateral $ABCD$ we have the following:

- $\measuredangle A \cong \measuredangle B$.
- $\measuredangle D \cong \measuredangle C$.
- $\measuredangle DNM \cong \measuredangle CNM$.
- $\measuredangle AMN \cong \measuredangle BMN$.
- $\overline{AD} \cong \overline{BC}$.
- $\overline{DN} \cong \overline{NC}$.
- $\overline{AM} \cong \overline{MB}$.
- $\overline{NM} \cong \overline{NM}$.

Therefore, $AMND$ is congruent to $BMNC$.

**5** First suppose that $ABCD$ is a parallelogram with four right angles. This means that $ABCD$ can be viewed as a Saccheri quadrilateral with any of its sides as its base. Now assume that $ABCD$ is both a Saccheri quadrilateral with base $\overline{AB}$ and base $\overline{AD}$. According to Euclid, since $\overline{AB}$ is parallel to $\overline{DC}$ and $\overline{AD}$ is parallel to $\overline{BC}$, we have that $ABCD$ is a parallelogram. Viewing $\overline{AB}$ as the base of $ABCD$, we see that $\measuredangle A$ and $\measuredangle B$ are right and $\measuredangle D \cong \measuredangle C$. When we view $\overline{AD}$ as its base, we conclude that $\measuredangle D$ is also right. Since $\measuredangle D \cong \measuredangle C$, we conclude that $\measuredangle C$ is a right angle.

**7(a)** Because $\overline{AD} \cong \overline{BC}$, $M$ is the midpoint of $\overline{AD}$, and $N$ is the midpoint of $\overline{BC}$, we have $\overline{MD} \cong \overline{NC}$. Moreover, $\overline{DC} \cong \overline{DC}$, and by Theorem 10.2.1, $\measuredangle MDC$ is congruent to $\measuredangle NCD$. So, $\triangle MDC \cong \triangle NCD$ by SAS. Hence, $\overline{MC} \cong \overline{DN}$.

**7(c)** Join $\overline{MN}$. We know that $\overline{MD} \cong \overline{NC}$. Because $ABNM$ is a Saccheri quadrilateral, its summit angles $\measuredangle AMN$ and $\measuredangle BNM$ are congruent. Thus, because we already know that $\measuredangle DMC \cong \measuredangle CND$ by the solution to Exercise 7(a), we conclude that $\measuredangle DMN$ is congruent to $\measuredangle CNM$ by angle addition, and then by SAS, $\triangle MDN \cong \triangle NCM$.

**9** Let $ABCD$ be a Saccheri quadrilateral. Assume the Parallel Postulate. This allows us to use Theorem 5.4.2. Since $\overline{AB}$ is parallel to $\overline{DC}$ [Exercise 3(b)], it must be the case that $\measuredangle A$ and $\measuredangle D$ are supplementary. Therefore, $\measuredangle D$ is a right angle, and because summit angles are congruent (Theorem 10.2.1), $\measuredangle C$ is also right.

**11** Here is the proof for Theorem 10.2.7. Assume **HR**$(ABCD)$. Let $EFGI$ be a Saccheri quadrilateral. Let $M$ be the midpoint of $\overline{EF}$ and $N$ the midpoint of $\overline{HG}$. Join $\overline{MN}$. We prove two of the three cases.

- If $AB = MN$, construct a Saccheri quadrilateral with base $\overline{MN}$ and side $\overline{EM}$. Its other side is a segment along $\overline{HN}$. Let $K$ be on $\overline{HN}$ (produced if necessary) such that $\overline{EM} \cong \overline{KN}$. If $KN < HN$, then $\overline{EK}$ is on the interior of $\measuredangle HEM$. Since **HR**$(ABCD)$, the summit angles of this Saccheri quadrilateral are right

(Lemma 10.2.6). Therefore, both $\sphericalangle HEM$ and $\sphericalangle KEM$ are right, which is impossible. We obtain a similar contradiction when $KN > HN$. Hence, $\overline{KN} \cong \overline{HN}$, which means that $K$ and $H$ are the same point. Therefore, $\sphericalangle EHG$ is a right angle, and then so it $\sphericalangle HGF$.

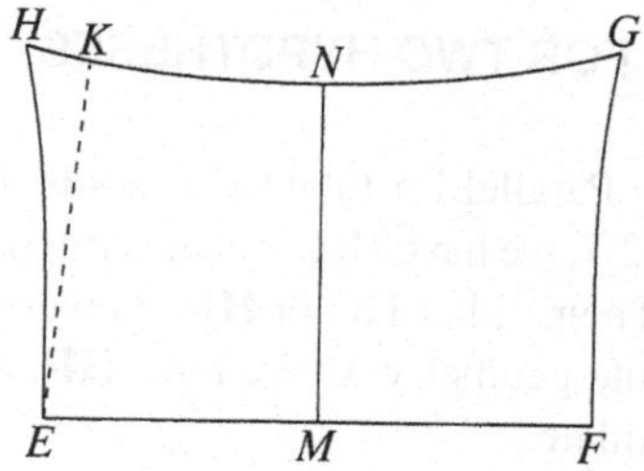


- Now suppose that $MN > AB$. Find $I$ on $\overline{MN}$ such that $\overline{MI} \cong \overline{AB}$. Find $K$ on $\overline{EH}$ so that $\overline{KI}$ is perpendicular to $\overline{IM}$. Note that if $KI > EM$, we can find $K'$ on $\overline{KI}$ (not pictured) such that $\overline{K'I} \cong \overline{EM}$, and then Saccheri quadrilateral $MIK'E$ has right summit angles. Hence, $\sphericalangle KEM$ and $\sphericalangle K'EM$ must be the same angle because each is a right angle. We obtain a similar situation when $KI < EM$. Therefore, $\overline{KI} \cong \overline{EM}$, and $MIKE$ is a Saccheri quadrilateral with $\sphericalangle EKI$ right.

  Now we may construct adjacent copies of $MIKE$ along $\overline{MN}$, creating a larger Saccheri quadrilateral, so we may assume that $MI > IN$. Find $J$ on $\overline{IM}$ such that $\overline{NI} \cong \overline{IJ}$. Erect a perpendicular to $\overline{MN}$ at $J$ intersecting $\overline{EK}$ at $L$. Then, $\overline{LJ} \cong \overline{KI}$; otherwise, we have a contradiction like in the previous paragraph. We should note that $\sphericalangle KLJ$ is a right angle because $\sphericalangle LKI$ is.

  Finally, join $\overline{KN}$ and $\overline{KJ}$. Then, $\Delta NIK \cong JIK$ by SAS. Therefore, $\sphericalangle NKI \cong \sphericalangle JKI$, which yields $\sphericalangle HKN \cong LKJ$. Similarly, $\sphericalangle HNK \cong LJK$. Hence, $\Delta HKN \cong \Delta LKJ$ by ASA, so $\sphericalangle EHN$ must be a right angle since it congruent to $\sphericalangle KLJ$. Hence, $\sphericalangle HGF$ is also a right angle because it is congruent to $\sphericalangle EHN$.

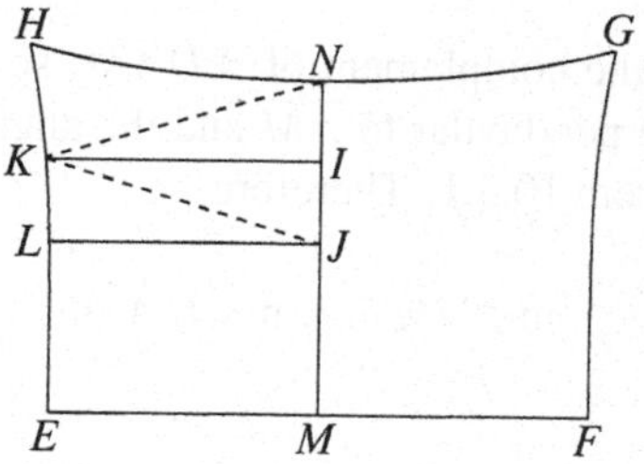


**13(a)** First note that since $m$ and $m'$ make the same angle with $t$, we may slide $m$ along $t$ (all the while keeping the angle that $m$ makes with $t$ fixed) until it is superimposed on $m'$. For this to happen, $m$ must cross $l$. Therefore, the point $C'$, which is an intersection of $m$ and $l$, exists as described in the problem. (What do you think about this argument?)

To complete the proof, appeal to Wallis's Postulate and find $C$ on the same side of $\overline{AB}$ as $C'$ such that $\triangle AB'C' \sim \triangle ABC$. Since $m\angle BAC = \alpha$ and $m\angle ABC = \beta$, $\overline{AC}$ must lie on $l$ and $\overline{BC}$ must lie on $m$. Hence, $l$ and $m$ intersect at $C$.

## 10.3 CONCLUSIONS FOR TWO HYPOTHESES

**1** First assume **HA**. If the Parallel Postulate also holds, then by Exercise 9 of Section 10.2 and Theorem 10.2.7, we have **HR**, which is impossible by Theorem 10.2.11.

Now assume not **HA**. Then, either **HO** or **HR**. However, by Theorem 10.3.5, **HO** is inconsistent with absolute geometry, so we have **HR**, and by Theorem 10.3.4 we also have the Parallel Postulate.

**3** Since **HR** implies the Parallel Postulate, quadrilateral drawn under the **HR** assumption are typical Euclidean quadrilaterals with straight sides. Since the angle sum must be less than four right angles under **HA**, the angles need to be pinched in:

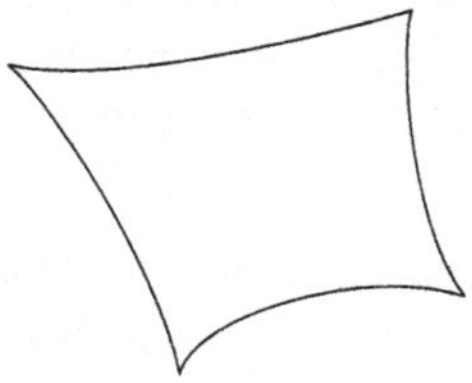

**5** Suppose **HR**. To prove $\overline{MD} \cong \overline{DF}$ we must take care not to use any result of Euclid that requires the Parallel Postulate. This is because Theorem 10.3.3 precedes Saccheri's proof that **HR** implies the Parallel Postulate. However, we can use Theorem 10.3.1. We will prove $\overline{MD} \cong \overline{DF}$ indirectly. First suppose that $MD > DF$. By the converse to Theorem 5.3.10,

$$m\angle DMF < m\angle DFM.$$

Observe that $\angle DFM$ is the complement of $\angle DAM$, so that $\angle DFM \cong \angle HAD$. This is because $\overline{CH}$ is perpendicular to $\overline{AM}$ and the angle sum of a triangle equals two right angles by Theorem 10.3.1. Therefore,

$$m\angle DMF < m\angle HAD,$$

and then we have

$$m\angle DMA > m\angle DAM$$

because $\angle DMA$ is a complement of $\angle DMF$ and $\angle DAM$ is a complement of $\angle HAD$. This is impossible because $MD > DF$ implies that $MD > AD$, so by the converse to Theorem 5.3.10, $m\angle DAM > m\angle DMA$. The case when $MD < DF$ is proven similarly.

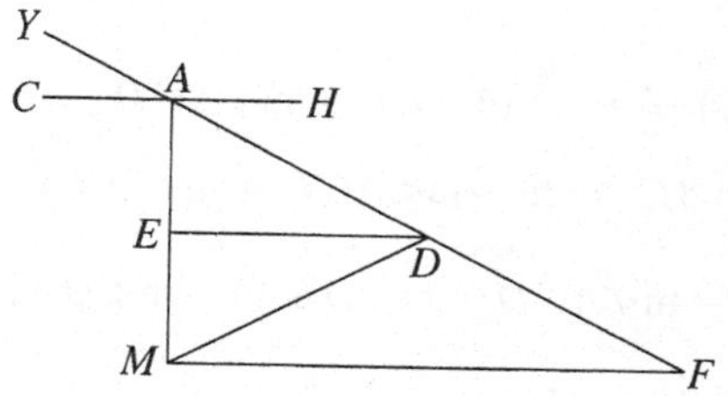


**7** We only need to make one modification to the proof of Theorem 10.3.4. Change the second displayed equation from

$$m\sphericalangle BAC + m\sphericalangle BCA = \text{a right angle}$$

to

$$m\sphericalangle BAC + m\sphericalangle BCA > \text{a right angle.}$$

We can still conclude that $\sphericalangle BAX$ is acute and apply Lemma 10.3.3.

## 10.4 PROPERTIES OF PARALLEL LINES

**1** Assume the following:

> If $\overline{AB}$ is perpendicular to $\overline{BY}$ with $X$ and $Y$ on the same side of $\overline{AB}$, then $\overline{AX}$ is parallel to $\overline{BY}$ if and only if $\sphericalangle BAX$ is right.

Let $\overline{BY}$ be given with $A$ not on $\overline{BY}$. Drop a perpendicular from $A$ to $\overline{BY}$. Assume that it is $\overline{AB}$. Find $X$ on the same side of $\overline{AB}$ as $Y$ such that $\overline{AX}$ is perpendicular to $\overline{AB}$. We know that $\overline{AX}$ is parallel to $\overline{BY}$ (Theorem 5.4.1). Any other line through $A$ will not be perpendicular to $\overline{AB}$, so by hypothesis such a line will intersect $\overline{BY}$. Thus, we have shown Playfair's Postulate.

Now suppose that we have Playfair's Postulate. Let $\overline{AB}$ be perpendicular to $\overline{BY}$. If we construct $\overline{AX}$ perpendicular to $\overline{AB}$, we know that it is parallel to $\overline{BY}$. Conversely, if a line through $A$ is parallel to $\overline{BY}$, it must be perpendicular to $\overline{AB}$ because Playfair's Postulate states that there is exactly one parallel to $\overline{BY}$ through $A$, namely the one that is perpendicular to $\overline{AB}$.

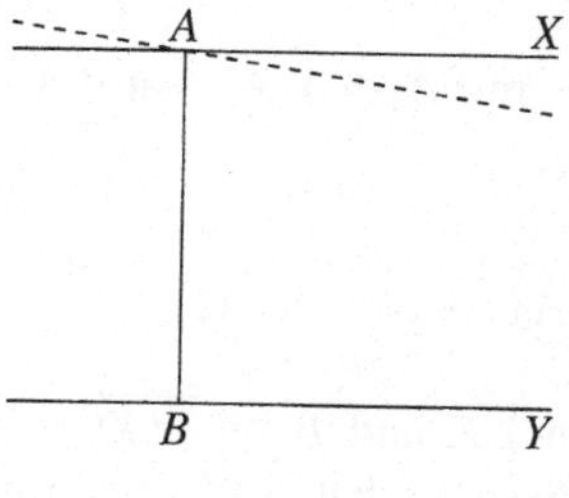


**3** In the interior of $\sphericalangle EAX$.

**5** In $\triangle ABC$ let $D$ be on $\overline{BC}$, forming $\triangle ABD$ and $\triangle ADC$. We know that

$$\mathsf{d}(ABD) = \pi - m\sphericalangle ADB - m\sphericalangle DBA - m\sphericalangle DAB$$

and

$$\mathsf{d}(ADC) = \pi - \text{m}\sphericalangle ADC - \text{m}\sphericalangle ACD - \text{m}\sphericalangle DAC.$$

Therefore, because $\mathsf{d}(ABC) = \pi - \text{m}\sphericalangle ACD - \text{m}\sphericalangle DBA - \text{m}\sphericalangle BAC$,

$$\mathsf{d}(ABC) = \pi - \text{m}\sphericalangle ACD - \text{m}\sphericalangle DBA - (\text{m}\sphericalangle DAC + \text{m}\sphericalangle DAB).$$

This implies that $\mathsf{d}(ABC)$ equals

$$\pi - \text{m}\sphericalangle ACD - \text{m}\sphericalangle DBA - \text{m}\sphericalangle DAC - \text{m}\sphericalangle DAB + \pi - \text{m}\sphericalangle ADC - \text{m}\sphericalangle ADB$$

because $\sphericalangle ADC$ and $\sphericalangle ADB$ are supplementary, and this last expression in turn equals

$$\pi - \text{m}\sphericalangle ADC - \text{m}\sphericalangle ACD - \text{m}\sphericalangle DAC + \pi - \text{m}\sphericalangle ADB - \text{m}\sphericalangle DBA - \text{m}\sphericalangle DAB.$$

Thus,

$$\mathsf{d}(ABC) = \mathsf{d}(ADC) + \mathsf{d}(ABD).$$

**6(a)** First we note that for convex polygon $A_1A_2\ldots A_n$,

$$\mathsf{d}(A_1A_2\ldots A_n) = \pi(n-2) - (\text{m}\sphericalangle A_1 + \text{m}\sphericalangle A_2 + \cdots + \text{m}\sphericalangle A_n).$$

Divide $P$ into $n-2$ triangles, each including $A_1$ as a vertex. Then,

$$\begin{aligned}
\mathsf{d}(A_1A_2A_3) &= \pi - \text{m}\sphericalangle A_1A_2A_3 - \text{m}\sphericalangle A_2A_3A_1 - \text{m}\sphericalangle A_3A_1A_2,\\
\mathsf{d}(A_1A_3A_4) &= \pi - \text{m}\sphericalangle A_1A_3A_4 - \text{m}\sphericalangle A_3A_4A_1 - \text{m}\sphericalangle A_4A_1A_3,\\
\mathsf{d}(A_1A_4A_5) &= \pi - \text{m}\sphericalangle A_1A_4A_5 - \text{m}\sphericalangle A_4A_5A_1 - \text{m}\sphericalangle A_5A_1A_4,\\
&\vdots\\
\mathsf{d}(A_1A_{n-1}A_n) &= \pi - \text{m}\sphericalangle A_1A_{n-1}A_n - \text{m}\sphericalangle A_{n-1}A_nA_1 - \text{m}\sphericalangle A_nA_1A_{n-1}.
\end{aligned}$$

Since

$$\text{m}\sphericalangle A_1 = \text{m}\sphericalangle A_3A_1A_2 + \text{m}\sphericalangle A_4A_1A_3 + \cdots + \text{m}\sphericalangle A_nA_1A_{n-1}$$

and

$$\text{m}\sphericalangle A_i = \text{m}\sphericalangle A_{i-1}A_iA_1 + \text{m}\sphericalangle A_1A_iA_{i+1}$$

for all $i = 3, 4, \ldots, n-1$,

$$\mathsf{d}(P) = \mathsf{d}(A_1A_2A_3) + \mathsf{d}(A_1A_3A_4) + \cdots + \mathsf{d}(A_1A_{n-1}A_n).$$

**7** Assume **HA**. Let $A$ be on $\overline{YX}$ and $B$ on $\overline{Y'X'}$ so that $\overline{AB}$ is perpendicular to $\overline{YX}$ and $\sphericalangle ABX'$ is acute. Assume that $\overline{BX'}$ does not intersect $\overline{AX'}$ produced but approaches it asymptotically. Let $E$ and $C$ be on $\overline{YA}$ with $C$ between $E$ and $A$. To obtain a contradiction, suppose that $\sphericalangle EFX'$ is not acute. Then by Theorem 10.3.2, $\sphericalangle FDC$ is not obtuse. Hence, $\sphericalangle BDC$ is not acute, so $\sphericalangle DBA$ is not obtuse. Finally, we have that $\sphericalangle ABX'$ is not acute, which is a contradiction. Therefore, $\sphericalangle EFX'$ is acute. Since $\overline{BX'}$ approaches $\overline{AX}$ asymptotically, $\overline{FX'}$ must also (Lemma 10.4.7), and we conclude that $FE > DC$.

**9** First note that by Equation 10.2 and SAS,

$$\Delta K_1K_2M \cong \Delta K_2K_3L.$$

From this we see that

$$\overline{K_1M} \cong \overline{K_2L}$$

and

$$\measuredangle NK_1M \cong \measuredangle HK_2L.$$

Therefore,

$$\Delta K_1NM \cong \Delta K_2HL.$$

From this we have

$$\measuredangle K_1NM \cong \measuredangle K_2HL.$$

This combined with Equation 10.3 shows that the angle sum of $K_2K_3LH$ is the angle sum of $K_1K_2MN$. This is the first observation. To prove the second, in quadrilateral $NMHD$, Theorem 10.3.2 implies that

$$\text{m}\measuredangle HDN + \text{m}\measuredangle DHM + (\pi - \text{m}\measuredangle MNK_1) + (\pi - \text{m}\measuredangle NMK_2) < 2\pi.$$

Therefore,

$$\text{m}\measuredangle MNK_1 + \text{m}\measuredangle NMK_2 > \text{m}\measuredangle HDN + \text{m}\measuredangle DHM.$$

When we add $\text{m}\measuredangle NK_1K_2 + \text{m}\measuredangle MK_2K_1$ to both sides of the inequality, we see that the angle sum of $K_1K_2MN$ is greater than $K_1K_2HD$.

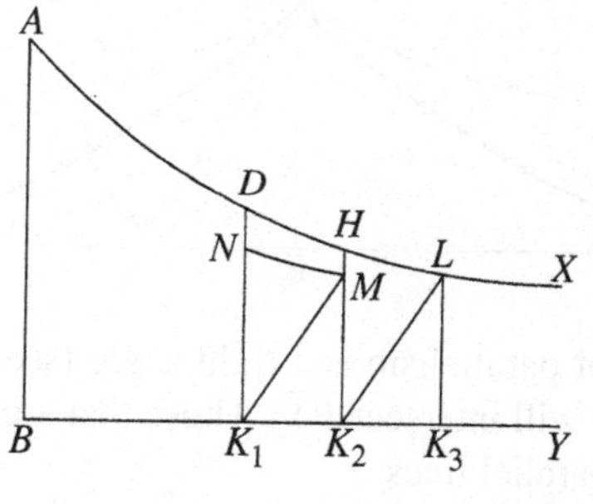


**11** Here is a method to approximate an asymptote. It requires the construction of perpendiculars and midpoints. Given $\overline{AB}$ and point $P$ such that $\overline{AP}$ is perpendicular to $\overline{AB}$, find $C_1, C_2, \ldots, C_n$ on $\overline{AB}$ such that

$$\overline{AC_1} \cong \overline{C_1C_2} \cong \cdots \cong \overline{C_{n-1}C_n}.$$

At each point erect perpendiculars $\overline{C_1D_1}, \overline{C_2D_2}, \ldots, \overline{C_nD_n}$ such that each $D_i$ is on the same side of $\overline{AB}$ as $P$ and

$$C_iD_i = 2C_{i+1}D_{i+1}.$$

Connecting the points $P$, $D_1$, $D_2$, ..., $D_n$, ... approximates an asymptote. A true asymptote cannot be constructed with straightedge and compass.

**13** He assumed that angle measure is continuous. The range of angle measure in the diagram for Theorem 10.4.2 is $(0, \pi)$.

**14** From calculus we have

$$\tan^{-1} x = x - \frac{x^3}{3} + \frac{x^5}{5} - \frac{x^7}{7} + \cdots + (-1)^n \frac{x^{2n+1}}{2n+1} + \cdots . \qquad (*)$$

Because $\tan^{-1} 1 = \pi/4$, equation $(*)$ yields

$$\frac{\pi}{4} = 1 - \frac{1}{3} + \frac{1}{5} - \frac{1}{7} + \cdots .$$

## 10.5 PARALLELISM REDEFINED

**1** Assuming **HA**, the situation is as shown in the diagram. Let $\overleftrightarrow{AN}$ be parallel to $\overleftrightarrow{BY}$. Lines like $\overleftrightarrow{AL}$ such that $m\sphericalangle BAL > m\sphericalangle BAN$ with $L$ on the same side of $\overleftrightarrow{AB}$ as $N$ will not intersect $\overleftrightarrow{BY}$. This means that $\overleftrightarrow{AL}$ is an example of a nonintersecting line. On the other hand, since $\overleftrightarrow{AK}$ is in the interior of $\sphericalangle BAN$, it is an example of an intersecting line. Notice that a mirror image of the right-hand side of $\overleftrightarrow{AB}$ is found on the left-hand side of $\overleftrightarrow{AB}$. In fact, we can then find a mirror image of the top side of $\overleftrightarrow{YY'}$ under $\overleftrightarrow{YY'}$.

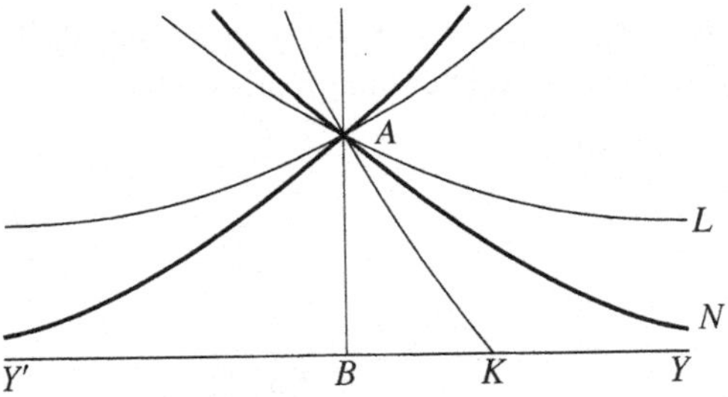


If **HR** holds, the angle of parallelism is a right angle (see Exercise 10), so all lines in the interior of $\sphericalangle BAN$ will intersect $\overrightarrow{BY}$. Those that are not will intersect $\overrightarrow{BY'}$. If **HO** holds, there are no parallel lines.

**3** Let $t$ be a transversal for $l$ and $m$. Let $A$ be the intersection of $t$ and $l$ and $B$ the intersection of $t$ and $m$. If $t$ is a common perpendicular, $m$ and $l$ are not parallel, so suppose that $t$ is not a common perpendicular. From this we conclude that $t$ makes an acute angle with one of the lines. Suppose it is $l$ and find $C$ on $l$ such that $\overleftrightarrow{BC}$ is perpendicular to $l$. To reach a contradiction, suppose that $l$ and $m$ are parallel. Letting $D$ be on $m$ as in the diagram, we know that $\sphericalangle CBD$ is acute. Find $E$ on the same side of $t$ as $D$ such that $\sphericalangle CBE$ is a right angle. Therefore,

$$m\sphericalangle ABC + m\sphericalangle CAB + m\sphericalangle BCA = m\sphericalangle ABC + m\sphericalangle CAB + m\sphericalangle CBD + m\sphericalangle DBE.$$

However, $m\sphericalangle ABC + m\sphericalangle CAB + m\sphericalangle CBD = \pi$ since the interior angles on the same side of $t$ are supplementary, so the sum of the internal angles of $\triangle ABC$ is greater than two right angles, a contradiction. Therefore, $l$ and $m$ cannot be parallel. (They are, however, nonintersecting.)

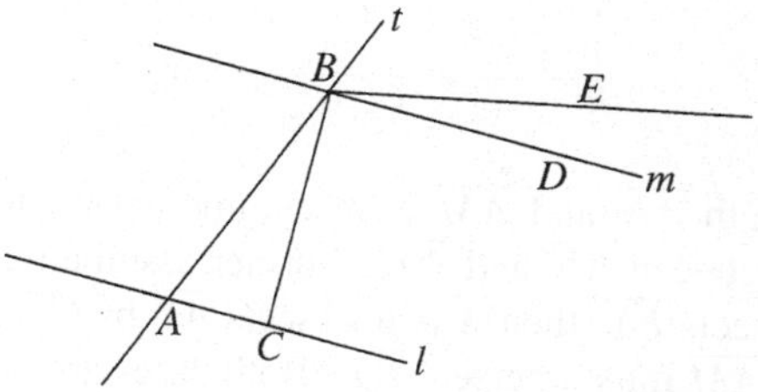


**4** Let $\overline{AB}$ be perpendicular to $\overline{BY}$ and $\overline{AX}$ parallel to $\overline{BY}$. Let $\overline{A'X'}$ and $\overline{B'Y'}$ be two other segments. Assume that $\sphericalangle BAX \cong \sphericalangle B'A'X'$, $\sphericalangle ABY \cong \sphericalangle A'B'Y'$, and $\overline{AB} \cong \overline{A'B'}$. Suppose that $\overline{A'X'}$ is not parallel to $\overline{B'Y'}$. There are two options.

- Let $E'$ be on $\overline{A'X'}$ and $F'$ on $\overline{B'Y'}$ so that $\overline{E'Y'}$ is a common perpendicular. Find $E$ on $\overline{AX}$ and $F$ on $\overline{BY}$ such that $\overline{AE} \cong \overline{A'E'}$ and $\overline{BF} \cong \overline{B'F'}$. This implies that $AEFB \cong A'E'F'B'$. Thus, $\overline{AX}$ and $\overline{BY}$ has a common perpendicular, a contradiction.

- The second case is when $\overline{A'X'}$ intersects $\overline{B'Y'}$. This leads to a similar contradiction.

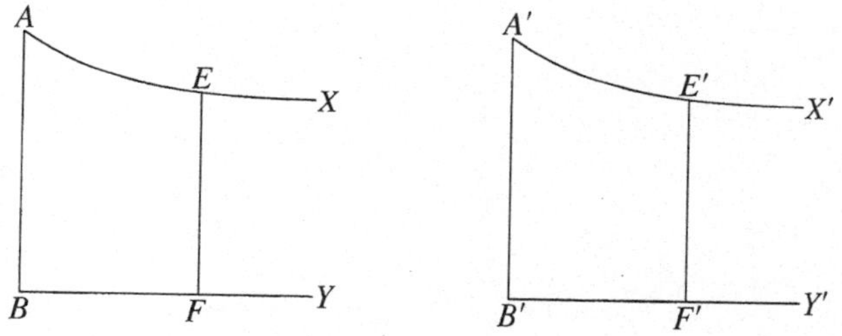


**7** The diagram below reflects the hypotheses of the problem under the assumption that $G'$ is between $B$ and $G$. Let $F'$ be the intersection of $\overline{GH}$ and $\overline{AX'}$ and let $F$ be the intersection of the $\overline{G'H'}$ produced and $\overline{AX}$. Now, $G'GF'H'$ has three right angles, so the fourth angle at $F'$ must be acute. This means that $\sphericalangle H'F'H$ is obtuse, giving $FHF'H'$ two right angles and an obtuse angle. Hence, $\sphericalangle H'FH$ must be acute, so $\sphericalangle H'FA$ is obtuse. Hence, the angle sum of $\triangle AH'F$ is greater than two right angles, contradicting **HA**. If $G$ and $G'$ are the same point, we arrive at a similar contradiction. Therefore, $G$ is between $B$ and $G'$.

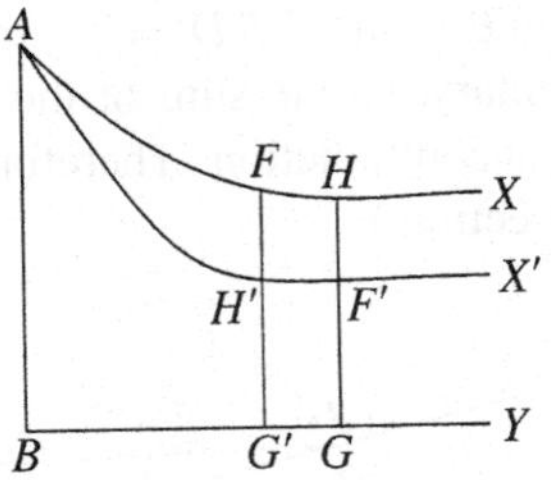


**9(b)** Suppose that both $\overline{AN}$ and $\overline{AM}$ have common perpendiculars with $\overline{BY}$ and suppose that $\overline{AL}$ is between $\overline{AN}$ and $\overline{AM}$. Further assume that $\overline{AM}$ is between $\overline{AL}$ and $\overline{BY}$. If $\overline{AL}$ intersects $\overline{BY}$, then $\overline{AM}$ intersects $\overline{AY}$ by Corollary 11.3.3. If $\overline{AL}$ is parallel to $\overline{BY}$, then $\overline{AM}$ must intersect $\overline{BY}$. Both cases are impossible. Hence, $\overline{AL}$ is a nonintersecting line and thus has a common perpendicular with $\overline{BY}$.

# CHAPTER 11

# JOHANN LAMBERT

## 11.1 THE THREE HYPOTHESES REVISITED

**1** On $\overline{AB}$, construct perpendiculars $\overline{AD}$ and $\overline{BC}$. At $D$, construct a perpendicular to $\overline{AD}$ that intersects $\overline{BC}$ at $C$. This completes the first Lambert quadrilateral. Extend $\overline{AB}$ beyond $A$ to $E$ so that $\overline{EA} \cong \overline{AB}$. Construct $\overline{EF}$ perpendicular to $\overline{EA}$ such that $\overline{EF} \cong \overline{BC}$. Join $\overline{DF}$. The Lambert quadrilateral $EADF$ is congruent to $ABCD$. For $EBCF$ to be a Saccheri quadrilateral, we need $F$, $D$, and $C$ to be collinear. For this, it suffices to show that $\sphericalangle FDA$ is a right angle (Theorem 5.3.8).

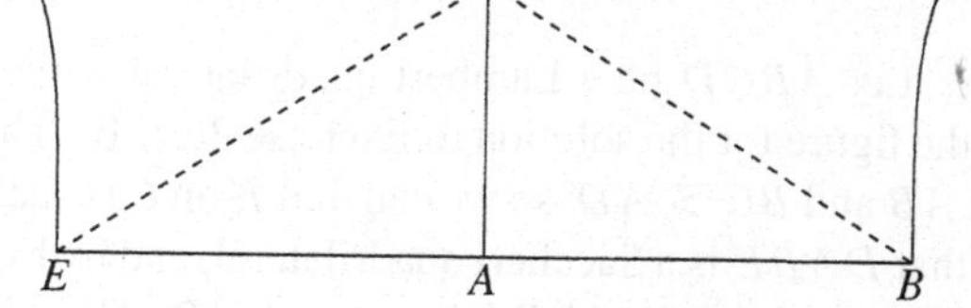


To prove this, join $\overline{ED}$ and $\overline{DB}$. By SAS, $\triangle EAD \cong \triangle BAD$. Therefore, $\sphericalangle EDA \cong \sphericalangle BDA$, $\sphericalangle DEA \cong \sphericalangle DBA$, and $\overline{ED} \cong \overline{DB}$. Since both $\sphericalangle FEA$

*Revolutions of Geometry.* By Michael O'Leary

and $\sphericalangle CBA$ are right, $\sphericalangle FED \cong \sphericalangle CBD$ by angle subtraction. Then by SAS, $\triangle EFD$ is congruent to $\triangle BCD$. Hence, $\sphericalangle FDE \cong \sphericalangle CDB$. By angle addition, $\sphericalangle FDA \cong \sphericalangle CDA$, and the latter angle is right.

**3(a)** In Lambert $ABCD$, assume that $\overline{AD} \cong \overline{BC}$ and $\overline{AB} \cong \overline{DC}$. Join $\overline{BD}$. Then $\triangle ABD \cong \triangle CDB$ by SSS. Hence, $\sphericalangle DAB \cong \sphericalangle DCB$, which means that the fourth angle of the Lambert quadrilateral is right. Hence, **HR** by Lemma 11.1.3.

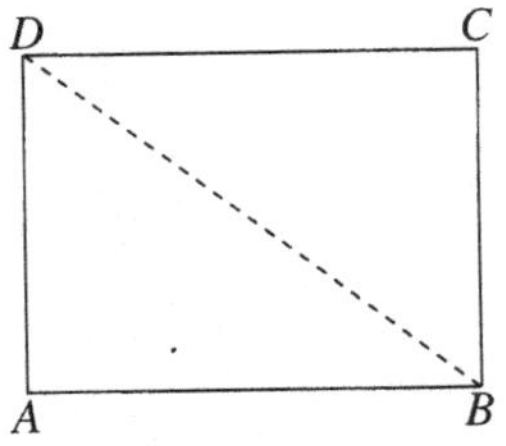


**3(c)** Let $ABCD$ be a Lambert quadrilateral. Suppose that $DC > AB$. Find $E$ on $\overline{DC}$ such that $\overline{AB} \cong \overline{DE}$. Join $\overline{EB}$. We now have Saccheri quadrilateral $DABE$ with base $\overline{AD}$. Since $\overline{EB}$ is on the interior of the right $\sphericalangle ABC$, $\sphericalangle ABE$ is acute. Thus, **HA**.

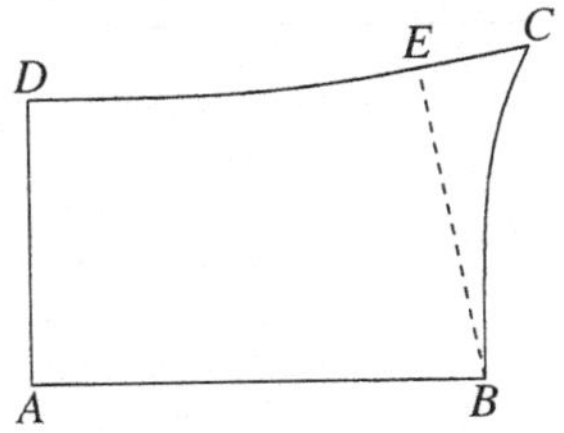


**5** The proof of HL depends on the Pythagorean Theorem, but the proof of the Pythagorean Theorem requires the Parallel Postulate. Therefore, HL cannot be appealed to if **HA** is assumed.

**7** Let $ABCD$ be a Saccheri quadrilateral with $M$ as the midpoint of $\overline{AB}$ and $N$ as the midpoint of $\overline{DC}$. Join $\overline{MN}$. Then, $\overline{MN}$ is a common perpendicular to $\overline{DC}$ and $\overline{AB}$ (Theorem 10.2.2). This gives us two Lambert quadrilaterals: $MADN$ and $MBCN$.

**9** We will prove **HA** (Saccheri version) if and only if the fourth angle of every Lambert quadrilateral is acute.

- Assume **HA**. Let $ABCD$ be a Lambert quadrilateral with $\sphericalangle C$ as the fourth angle, as in the figure for the solution to Exercise 3(c). By Theorem 11.1.1, we have $CD > AB$ and $BC > AD$, so we can find $E$ on $\overline{CD}$ such that $\overline{DE} \cong \overline{AB}$. This means that $DABE$ is a Saccheri quadrilateral, and by hypothesis, $\sphericalangle DEB$ is acute. By Theorem 5.3.9, m$\sphericalangle DEB >$ m$\sphericalangle ECB$. Therefore, $\sphericalangle ECB$ must be acute.

- Now suppose that the fourth angle of every Lambert quadrilateral is acute. Let $ABCD$ be a Saccheri quadrilateral with base $\overline{AB}$. Let $M$ and $N$ be the midpoints of $\overline{AB}$ and $\overline{DC}$, respectively. This results in two Lambert quadrilaterals such that the fourth angles of these quadrilaterals are the summit angles of the Saccheri quadrilateral. Therefore, the summit angles are acute, and we have **HA**.

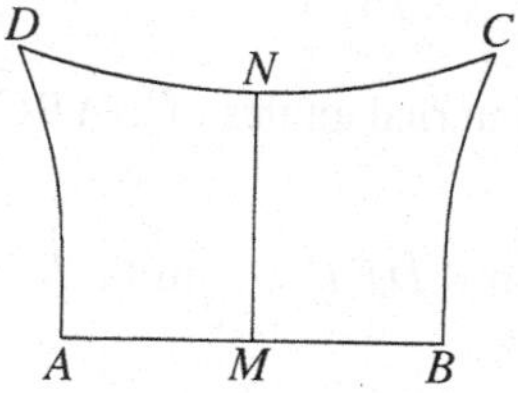


## 11.2 POLYGONS

**1** The answer to this question depends on which of the three hypotheses holds. If **HR** is true, there is no difference between a rectangle, a Saccheri quadrilateral, and a Lambert quadrilateral. If **HA** or **HO** hold true, there are a few similarities between the figures other than the fact that they are quadrilaterals and some of their angles are right. Of course, this last sentence refers to the commonalities between a Saccheri quadrilateral and a Lambert quadrilateral. This is because under **HA** and **HO**, there are no rectangles.

**3** Suppose that $ABCD$ is both a Saccheri and a Lambert quadrilateral. Let $\overline{AB}$ be its base as a Saccheri quadrilateral. This means that one of the summit angles will serve as the fourth angle of the Lambert quadrilateral. Suppose that it is $\sphericalangle C$. Hence, $\sphericalangle D$ is right because $ABCD$ is Lambert. Since the summit angles of a Saccheri quadrilateral are congruent, $\sphericalangle C$ must be right. Hence, $ABCD$ is a rectangle (and thus **HR** holds).

**5(a)** We are given $\triangle ABC$. Let $M$ be the midpoint of $\overline{AC}$. Join $\overline{BM}$ and extend it to $G$ so that $M$ is the midpoint of $\overline{BG}$. Join $\overline{CG}$. Since $\sphericalangle AMB \cong \sphericalangle GMC$, we conclude that $\triangle AMB \cong \triangle CMG$ by SAS. Hence,

$$\sphericalangle BAC \cong \sphericalangle ACG \text{ and } \sphericalangle ABG \cong \sphericalangle BGC.$$

So,

$$m\sphericalangle ABC + m\sphericalangle BCA + m\sphericalangle BAC$$

equals

$$m\sphericalangle ABM + m\sphericalangle CBG + m\sphericalangle BCA + m\sphericalangle ACG,$$

but this in turn equals

$$m\sphericalangle BGC + m\sphericalangle CBG + m\sphericalangle BCA + m\sphericalangle ACG.$$

Therefore, the angle sum of $\triangle ABC$ equals the angle sum of $\triangle BGC$. We now have one of two cases:

- If $m\sphericalangle CBG \leqslant \frac{1}{2} m\sphericalangle ABC$, construct $\triangle DEF$ so that $\triangle DEF \cong \triangle CBG$.

- If $m\sphericalangle ABG \leqslant \frac{1}{2} m\sphericalangle ABC$, then $m\sphericalangle BGC \leqslant \frac{1}{2} m\sphericalangle ABC$. In this case construct $\triangle DEF$ so that $\triangle DEF \cong \triangle BGC$.

In both cases, the sum of the internal angles of $\triangle ABC$ equals the sum of the internal angles of $\triangle DEF$, and

$$m\sphericalangle DEF \leqslant \frac{1}{2} m\sphericalangle ABC.$$

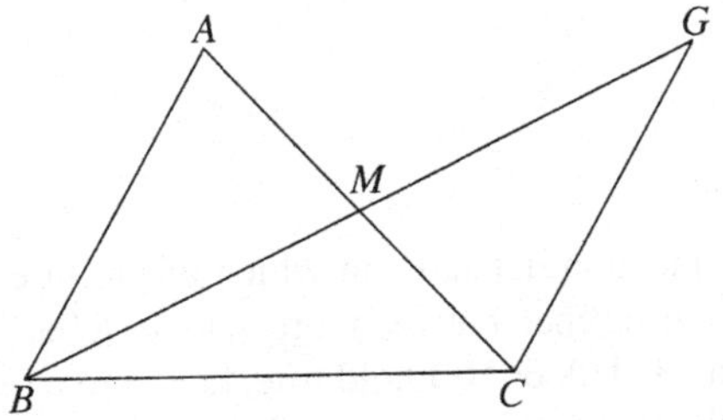


**5(c)** It appears so, because in the proof of Exercise 5(a) we see that $BG$ is greater than all of the sides of $\triangle ABC$. As Legendre refers continually to Exercise 5(a) in his proof of the Saccheri-Legendre Theorem, the sequence of triangles give rise to a sequence of segments with lengths that grow without bound. Therefore, we can at least conclude that Legendre assumed that the possible lengths of lines are unbounded, and, furthermore, it appears safe to assume that he (as did Euclid) probably viewed his lines as infinite in length.

**7** Let $\sphericalangle ABC$ be external to $\triangle BDC$. We know that

$$m\sphericalangle ABD = \pi - m\sphericalangle DBC.$$

By the Saccheri-Legendre Theorem (11.2.7),

$$m\sphericalangle DBC + m\sphericalangle BCD + m\sphericalangle BDC \leqslant \pi,$$

so

$$m\sphericalangle DBC - \pi + m\sphericalangle BCD + m\sphericalangle BDC \leqslant 0.$$

Therefore,

$$m\sphericalangle BCD + m\sphericalangle BDC - m\sphericalangle ABD \leqslant 0,$$

from which we obtain

$$m\sphericalangle ABD \geqslant m\sphericalangle BCD + m\sphericalangle BDC.$$

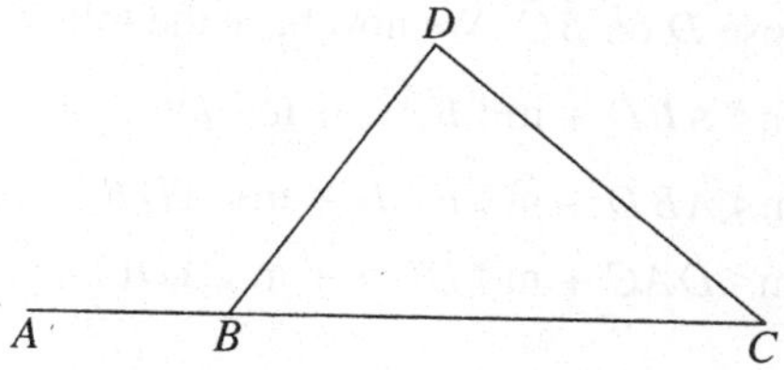


**9** Take $n > 2$. Let $A_1A_2 \ldots A_n$ be a convex polygon with $n + 1$ sides. This polygon is the union of

$$\triangle A_1A_2A_3,\ \triangle A_1A_3A_4,\ \triangle A_1A_4A_5,\ \ldots,\ \triangle A_1A_{n-1}A_n.$$

There are $n - 2$ triangles in this list. Since the angle sum of the polygon is the sum of the angle sums of each of the triangles, we conclude by the Saccheri-Legendre Theorem (11.2.7) that the angle sum of the convex polygon is no more than $n - 2$ right angles.

**11** If there exists a triangle with zero defect, there exists a rectangle.

**13** In absolute geometry we have the following:

- Assume that all triangles have angle sum less than two right angles. Therefore, rectangles do not exist because if they did, we could construct a triangle with angle sum equal to two right angles. To see this, suppose that $ABCD$ is a rectangle. This means that it can be viewed as a Lambert quadrilateral with a right fourth angle. Hence, opposite sides are congruent by Theorem 11.1.1. Join $\overline{DB}$. Then $\triangle ABD \cong \triangle CDB$ by SSS. Thus, $\sphericalangle ABD \cong \sphericalangle BDC$, from which we conclude that the angle sum of $\triangle ABD$ equals two right angles because

$$\text{m}\sphericalangle ADB + \text{m}\sphericalangle DAB + \text{m}\sphericalangle ABD = \text{m}\sphericalangle ADB + \text{m}\sphericalangle DAB + \text{m}\sphericalangle BDC = \pi.$$

  The last equality holds because $\sphericalangle DAB$ is a right angle and $\sphericalangle ADB$ and $\sphericalangle BDC$ are complementary. We may now conclude that since rectangles do not exist, the Parallel Postulate is false (Theorem 11.2.2), so we have **HA** (Corollary 11.2.14).

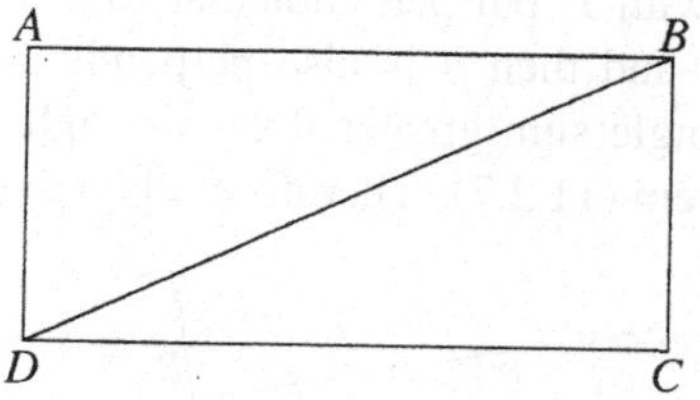


- Let **HA** hold. Therefore, **HR** is false, and then the Parallel Postulate does not hold (Theorem 10.3.4). Hence, by Theorem 11.2.2, rectangles do not exist, so by Theorems 11.2.10 and 11.2.7, all triangles have angle sum less than two right angles.

**15** Take $\triangle ABC$. Choose $D$ on $\overline{BC}$. We now have the following:

$$m\measuredangle ABD + m\measuredangle BAC + m\measuredangle DCA = n,$$
$$m\measuredangle ABD + m\measuredangle BAD + m\measuredangle ADB = n,$$
$$m\measuredangle DAC + m\measuredangle DCA + m\measuredangle ADC = n,$$

and

$$m\measuredangle ADB + m\measuredangle ADC = \pi.$$

Therefore,

$$m\measuredangle ABD + m\measuredangle BAD + m\measuredangle DAC + m\measuredangle DCA + m\measuredangle ADB + m\measuredangle ADC = 2n,$$

and then

$$n + \pi = 2n.$$

Hence, $n = \pi$, and we conclude that the internal angle sum of $\triangle ABC$ equals two right angles, which implies that the Parallel Postulate holds (Corollary 11.2.12).

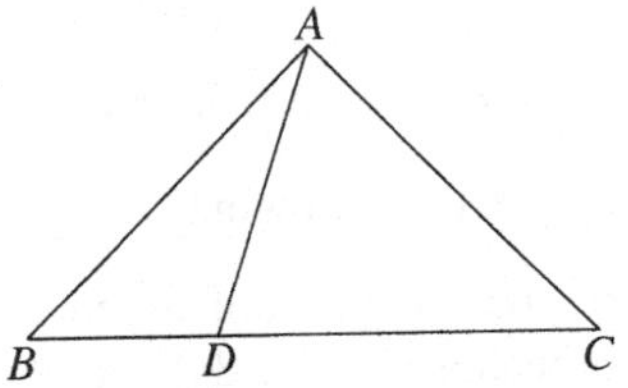


**16(a)** Assume the Parallel Postulate. Suppose that $l$ and $m$ are parallel and that $n$ is perpendicular to $l$ at $P$. Notice that $n$ must intersect $m$; otherwise, we would have two lines through $P$ that are parallel to $m$, contradicting Playfair's Postulate (10.1.1), which is logically equivalent to the Parallel Postulate (Theorem 10.1.2). This means that $n$ is a transversal for the parallel lines $l$ and $m$. Hence, by Theorem 5.4.2, interior angles on the same side of $n$ are supplementary, so $n$ makes a right angle with $m$.

Now suppose that

if $l$ and $m$ are parallel and $n$ is perpendicular to $l$,
$n$ is perpendicular to $m$.

To prove that the Parallel Postulate follows, we will show Playfair's Postulate indirectly. Assume that $l$ and $s$ are distinct lines through $P$ that are parallel to $m$. Let $n$ be a line not incident with $P$ but perpendicular to $l$ at $A$. Then, by hypothesis, $n$ is perpendicular to $m$, and then $n$ is also perpendicular to $s$ at some point $B$. Therefore, $\triangle ABP$ has angle sum greater than two right angles, contradicting the Saccheri-Legendre Theorem (11.2.7). Therefore, Playfair's Postulate holds.

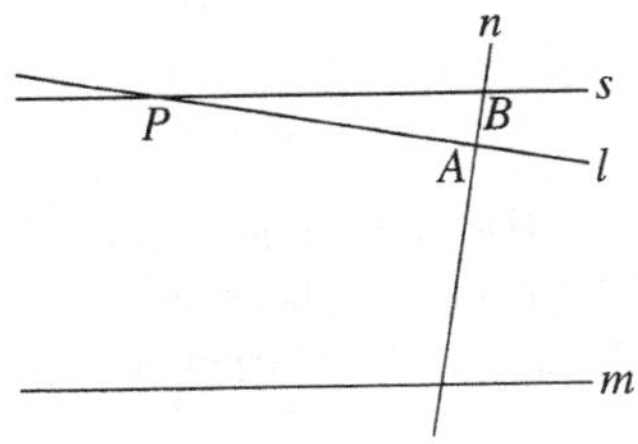

**16(c)** That the Parallel Postulate implies the proposition follows from Exercise 11 of Section 5.5. To prove the converse, assume that

the perpendicular bisectors of the sides of a triangle are concurrent. (∗)

To prove the Parallel Postulate, it suffices to show that Exercise 16(b) follows from (∗). If we examine the proof of Theorem 5.5.8 we see that (∗) implies that we can circumscribe a circle about a triangle. Therefore, we conclude that

there exists a point that is equidistant from any three noncollinear points. (†)

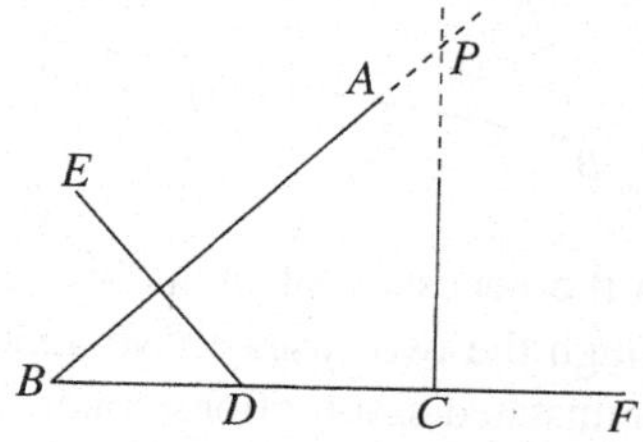


We use this fact as follows: Let $\measuredangle ABC$ be an acute angle. Let $D$ be the midpoint of $\overline{BC}$ and find $E$ on the opposite side of $\overline{AB}$ as $D$ such that $\overline{AB}$ is the perpendicular bisector of $\overline{DE}$. Choose $F$ on the $\overline{BC}$ produced such that $\overline{DC} \cong \overline{CF}$. Observe that $E$, $D$, and $F$ are not collinear, so by (†) there exists a point $P$ that is equidistant between $D$, $E$, and $F$. Since $P$ is equidistant between $D$ and $E$, it must lie on $\overline{AB}$ (produced), and since $P$ is equidistant between $D$ and $F$, it must lie on the line perpendicular to $\overline{BC}$ at $C$. Thus, we have proven the proposition of Exercise 16(b).

## 11.3 OMEGA TRIANGLES

**1** This sentence is the same as Pasch's Postulate except that it does not have the uniqueness condition. Take $\triangle ABC$. Suppose that a line $l$ enters the triangle through $\overline{AB}$ at $D$, which is not a vertex. This means that there exists $E$ on the interior of $\triangle ABC$ such that $E$ is on $l$. By hypothesis, $l$ intersects one of the other sides, say $\overline{AC}$, at $F$. To reach a contradiction, assume that $l$ also intersects $\overline{BC}$ at $G$. Note that none of these new points are vertices. Furthermore, $l$ cannot intersect the same side twice because then two points would determine two lines. We have two cases:

- $l$ exits the triangle at $F$ and returns to intersect $\overline{BC}$ at $G$. In this case we find that $\triangle CFG$ has angle sum greater than two right angles, contradicting the Saccheri-Legendre Theorem (11.2.7).

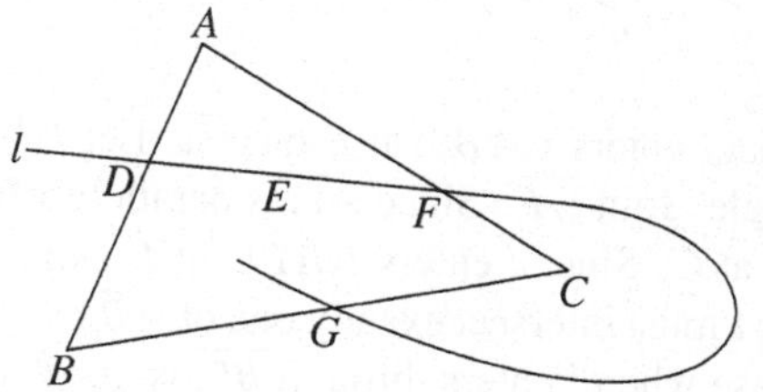

- $l$ does not exit the triangle at $F$, yet returns to meet $\overline{BC}$ at $G$. There are multiple problems with this. First, there is an unstated assumption (probably dating to before Euclid) that if two lines intersect, they cross. Also, since $E$ is on the interior of the triangle, we should be able to join $C$ and $E$ to form a segment that is not on $l$. However, this leads to $l$ intersecting $\overline{EC}$ twice.

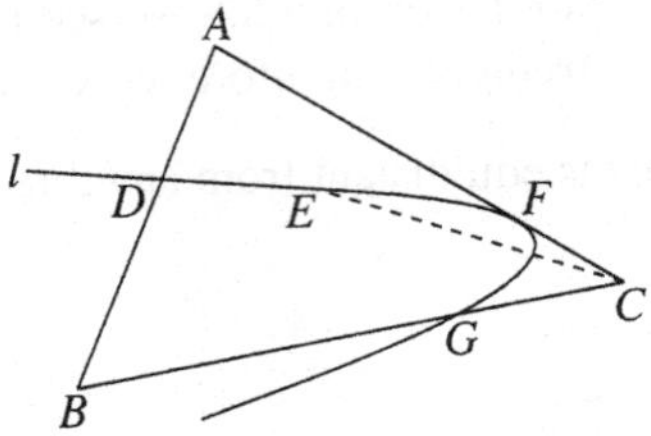


This exercise demonstrates the necessity of including uniqueness in the statement of Pasch's Postulate. Although the two cases above appear to yield contradictions, these contradictions rely on unstated assumptions concerning the nature of lines.

**3** We must prove

> If a line enters an omega triangle through a side and does not contain one of its vertices, the line intersects exactly one of the other sides of the triangle.

Given $\triangle AB\Omega$, suppose that $l$ enters the triangle. We have two cases to consider.

- Let $l$ intersect $\overline{AB}$ at $E$ that is not $A$ and not $B$. This means that there exists a point $F$ on the interior of $\triangle AB\Omega$ that is on $l$. Notice that $F$ is not on $\overline{E\Omega}$; otherwise, $l$ would contain the vertex $\Omega$. Suppose that $F$ is in the interior of $\triangle AE\Omega$, so $l$ enters $\triangle AE\Omega$ at vertex $E$. Therefore, $l$ must intersect $\overline{A\Omega}$ at some $D$ by Theorem 11.3.4. If $l$ also intersects $\overline{B\Omega}$ at some point $C$, points $C$, $E$, and $D$ are collinear, so $C$ and $D$ are on two different lines, which is impossible. The case when $F$ is in the interior of $\triangle EB\Omega$ is proven similarly.

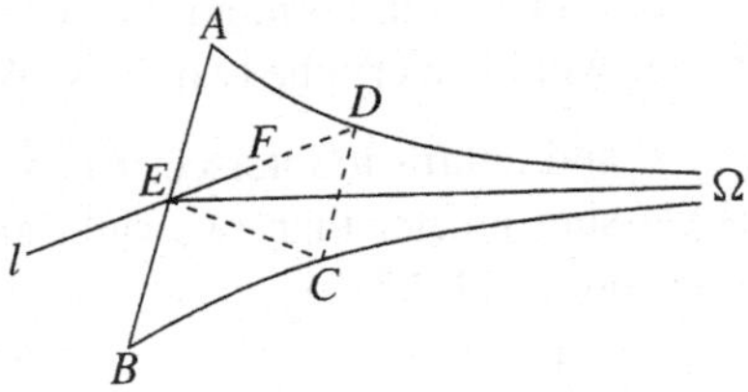


- Now suppose that $l$ enters $\triangle AB\Omega$ at $E$ on $\overline{A\Omega}$. Let $F$ be on $l$ in the interior of the omega triangle. Join $\overline{AF}$. Since $\overline{A\Omega}$ is parallel to $\overline{B\Omega}$, we can extend $\overline{AF}$ to intersect $\overline{B\Omega}$ at $C$. Since $l$ enters $\triangle ABC$ at $F$ and $l$ contains no vertices of $\triangle ABC$, the line $l$ must intersect exactly one of $\overline{AB}$ or $\overline{BC}$ by Pasch's Postulate (11.3.2). The case when $l$ enters through $\overline{B\Omega}$ is handled in the same manner.

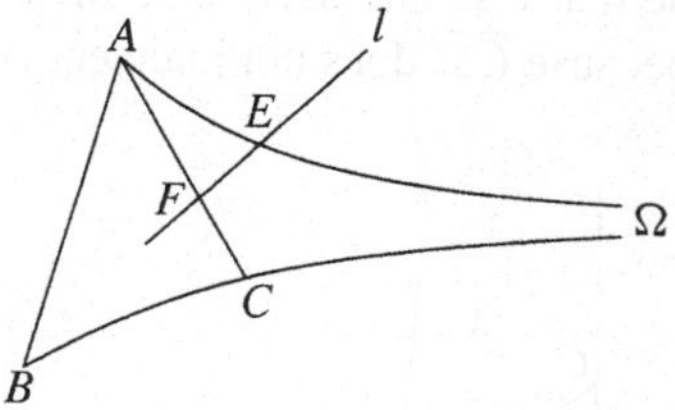


**5** The proposition that must be true is this:

there exists exactly one constant of proportionality for the areas of all triangles.

Assume this statement and let $k$ be the constant of proportionality. Let $ABCD$ be a quadrilateral that is the disjoint union of $\triangle ABC$ and $\triangle ADC$. We know that

$$\mathsf{d}(ABCD) = \mathsf{d}(ABC) + \mathsf{d}(ADC),$$

$$\text{area}(\triangle ABC) = k\mathsf{d}(ABC),$$

and

$$\text{area}(\triangle ACD) = k\mathsf{d}(ACD).$$

Therefore,

$$\begin{aligned}\text{area}(ABCD) &= \text{area}(\triangle ABC) + \text{area}(\triangle ACD)\\ &= k[\mathsf{d}(ABC) + \mathsf{d}(ACD)]\\ &= k\mathsf{d}(ABCD).\end{aligned}$$

**7** In $\triangle AB\Omega$, let $M$ be the midpoint of $\overline{AB}$ and $\sphericalangle AB\Omega \cong \sphericalangle BA\Omega$. Then, $\triangle AM\Omega$ is congruent to $\triangle BM\Omega$ by SA. Therefore, $\sphericalangle AM\Omega \cong \sphericalangle BM\Omega$, which implies that $\overline{M\Omega}$ is perpendicular to $\overline{AB}$.

**9** When we draw the pictures, we must keep in mind the angle of parallelism (page 355 of the text). Here are two omega triangles with $\overline{AB} \cong \overline{A'B'}$. In the diagrams, $\overline{AB}$ is perpendicular to $\overline{B\Omega}$ and $M$ is the midpoint of $\overline{A'B'}$. These triangles are not congruent because $\triangle AB\Omega$ is a right omega triangle, but $\triangle A'B'\Omega'$ is not. Instead, both $\sphericalangle MA'\Omega'$ and $\sphericalangle MB'\Omega'$ are acute.

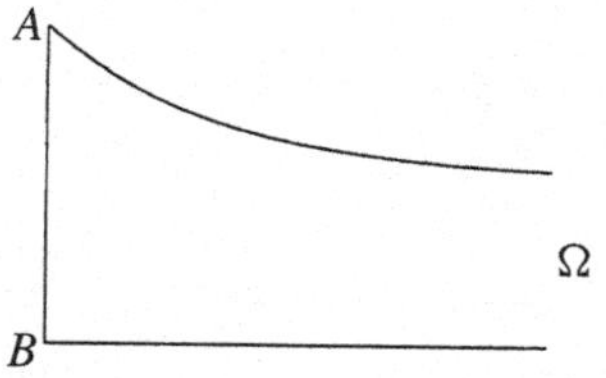


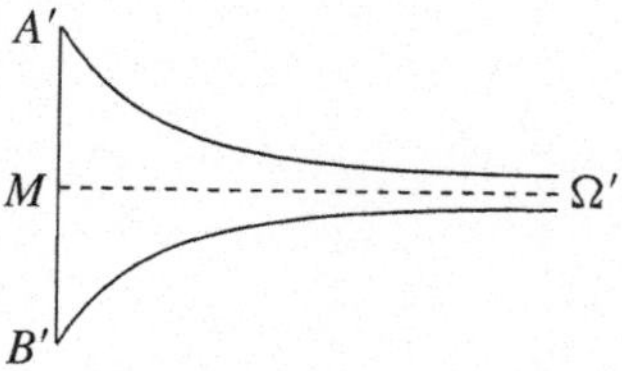


**11(a)** Let $B$ be on $\overline{A\Omega}$ and $C$ on $\overline{A\Omega'}$. Assume that $\overline{A\Omega}$ is perpendicular to $\overline{A\Omega'}$. Notice that $\overline{C\Omega}$ is parallel to $\overline{A\Omega}$ and $\overline{B\Omega'}$ is parallel to $\overline{A\Omega'}$. Because $\overline{C\Omega}$ enters

$\triangle AB\Omega'$ at $C$, we conclude that $\overline{C\Omega}$ intersects $\overline{B\Omega'}$ at some $D$ by Theorem 11.3.5. We draw this conclusion because $\overline{C\Omega}$ does not intersect $\overline{A\Omega}$.

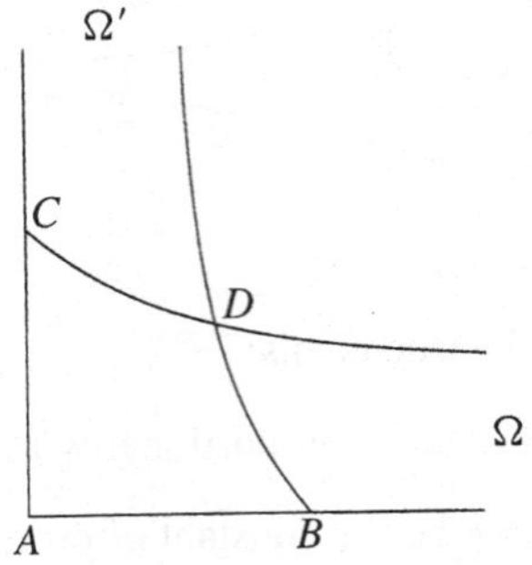


**13** No, because the Pythagorean Theorem is equivalent to the Parallel Postulate.

## 11.4 PURE REASON

The are no exercises for this section.

# CHAPTER 12

# NICOLAI LOBACHEVSKI AND JÁNOS BOLYAI

## 12.1 PARALLEL FUNDAMENTALS

**1** No.

**3** Suppose that $\overline{AX}$ is parallel to $\overline{BY}$. This means that there exists ideal point $\Omega$ such that $\Omega$ is on $\overline{AX}$ and $\overline{BY}$. Since the two lines share the same ideal point, $\overline{BY}$ is parallel to $\overline{AX}$.

**5** Let $l$ and $m$ be two lines in space. We say that $l$ and $m$ are **parallel** if there exists a plane $\Lambda$ such that $l$ and $m$ are both in $\Lambda$ and are parallel in this plane.

**7** Since

$$\Pi : (0, \infty) \to [0, \pi/2)$$

is a strictly decreasing, continuous, one-to-one correspondence (Theorems 12.1.4–12.1.7),

$$\lim_{p \to 0+} \Pi(p) = \frac{\pi}{2}.$$

**9** True. Here is the proof:

- Suppose not **HR**. By the Saccheri-Legendre Theorem (11.2.7), we may conclude **HA**. Hence, since $\Pi$ is a strictly decreasing function (Theorem 12.1.4) and
$$\lim_{p \to 0+} \Pi(p) = \frac{\pi}{2}$$
(Exercise 7), there cannot exist $p > 0$ such that $\Pi(p) = \pi/2$.

- Assume **HR**. Let $\overline{AB}$ be perpendicular to both $\overline{AX}$ and $\overline{BY}$, with $X$ and $Y$ on the same side of $\overline{AB}$. Let $\overline{AD}$ be in the interior of $\sphericalangle BAX$. Since **HR** is equivalent to Playfair's Postulate (Theorem 11.2.19), $\overline{AD}$ must intersect $\overline{BY}$. Therefore, $\overline{AX}$ is parallel to $\overline{BY}$, and $\Pi(AB) = \pi/2$.

**11** Let $A$ not be on $l$. Let $\overline{AN}$ and $\overline{AN'}$ both be parallel to $l$. Let $\overline{AB}$ be the angle bisector of $\sphericalangle N'AN$. Since $\overline{AN}$ and $\overline{AN'}$ are parallel to $l$, we may assume that $B$ is on $l$. Now let $\Omega$ and $\Omega'$ be ideal points such that $\Omega$ is on $\overline{AN}$ and $\Omega'$ is on $\overline{AN'}$. Because $\sphericalangle BA\Omega' \cong \sphericalangle BA\Omega$ and $\overline{AB} \cong \overline{AB}$, we have $\Delta AB\Omega \cong \Delta AB\Omega'$ by SA. Thus, $\sphericalangle AB\Omega' \cong \sphericalangle AB\Omega$, which implies that $\overline{AB}$ is perpendicular to $l$.

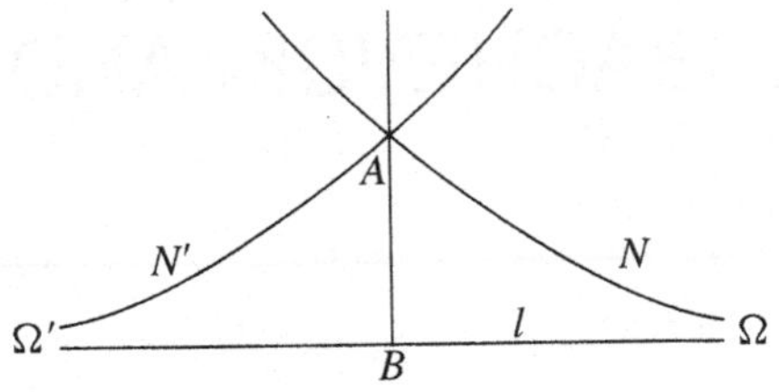


## 12.2 HOROCYCLES

**1** This is Proposition 5.4 on page 167 of the text.

**3** Let $\overline{AB}$ be parallel to $\overline{CD}$, $\sphericalangle CAB \cong \sphericalangle ACD$, and $\overline{EF}$ be the perpendicular bisector of $\overline{AC}$. Therefore, $\overline{EF}$ is the line of symmetry between $\overline{AB}$ and $\overline{CD}$ (Theorem 12.2.2). Because it is the line of symmetry, $\overline{EF}$ is equidistant between $\overline{AB}$ and $\overline{CD}$, so it does not intersect either of these parallel lines. To show that $\overline{EF}$ is parallel to $\overline{AB}$, we will show that $\overline{AB}$ is parallel to $\overline{EF}$ (Theorem 12.1.2). Let $\overline{AG}$ lie in the interior of $\sphericalangle CAB$. Because $\overline{AB}$ is parallel to $\overline{CD}$, $\overline{AG}$ intersects $\overline{CD}$. Assume that $G$ is the point of intersection. This forms $\Delta ACG$ with $\overline{EF}$ entering this triangle through the side $\overline{AC}$. Therefore, by Pasch's Postulate (11.3.2), $\overline{EF}$ intersects $\overline{AG}$, so $\overline{AB}$ is parallel to $\overline{EF}$, and then $\overline{EF}$ is also parallel to $\overline{CD}$ (Theorem 12.1.3).

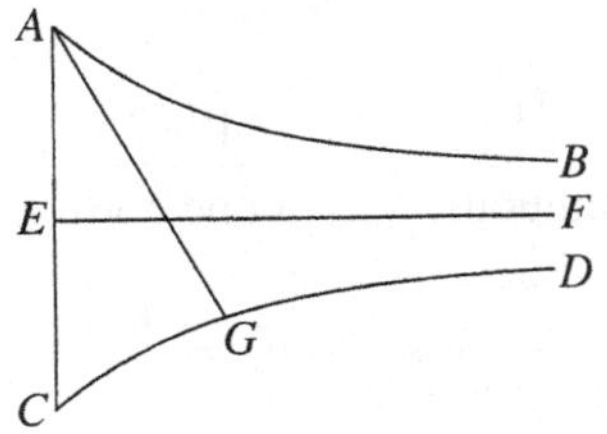

**5** Assuming **HR**, a horocycle is a line, so in this case between two points there is exactly one horocycle. Suppose **HA**, instead. Given two points in the plane, there will be two horocycles through the two points. This is illustrated in the diagram by horocycles $h$ and $i$.

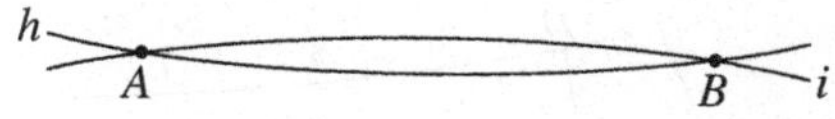


**7** See Section 13.4.

**9** A horocycle has constant curvature, so the same surface results from rotating a horocycle about any of its axes. Since any two horocycles have the same curvature, the surfaces that result from rotating these horocycles about their axes can be considered to be congruent. To confirm that a horocycle has constant curvature, we can measure the curvature as follows: Let $\mathscr{L}$ be a horocycle and let $\Omega$ be the ideal point through which the axes of $\mathscr{L}$ pass. Take $A$, $B$, $C$, and $D$ on $\mathscr{L}$ such that $\overline{AB} \cong \overline{CD}$. To show that the curvature of $\mathscr{L}$ is constant, we will show that $\sphericalangle BA\Omega \cong \sphericalangle DC\Omega$. Let $M$ be the midpoint of $\overline{AB}$ and $N$ the midpoint of $\overline{CD}$. Hence, $\overline{AM} \cong \overline{CN}$, so by SA we have that $\triangle AM\Omega \cong \triangle CN\Omega$. Therefore, $\sphericalangle BA\Omega \cong \sphericalangle DC\Omega$. Furthermore, if $\mathscr{L}'$ was another horocycle, we can use the same strategy to show that $\mathscr{L}'$ has the same curvature of $\mathscr{L}$.

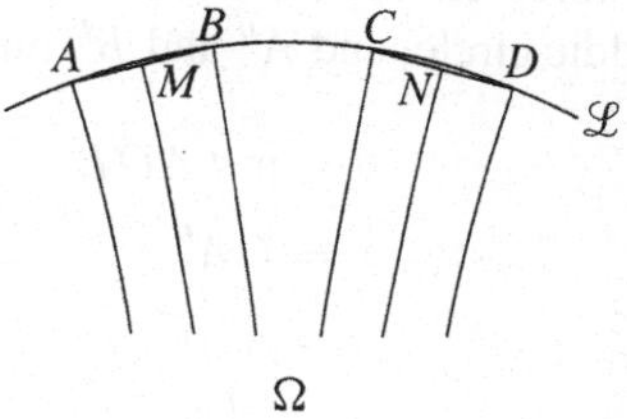


**11** First construct the bisector of $\sphericalangle BA\Omega$. It intersects $\overline{B\Omega}$ at $H$ because $\overline{A\Omega}$ is parallel to $\overline{B\Omega}$. Next construct the angle bisector of $\sphericalangle AB\Omega$. This bisector enters $\triangle ABH$ at $B$, so it intersects $\overline{AH}$ at some $E$ (Theorem 11.3.3). Notice that it is not important that the two lines bisect the angles. What is important is that $\overline{AH}$ is in the interior of $\sphericalangle BA\Omega$ and $\overline{BE}$ is in the interior of $\sphericalangle AB\Omega$.

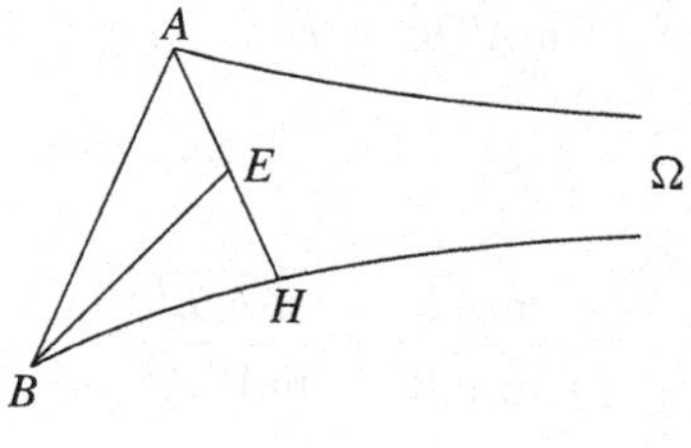


**13(a)** Let $n$ be the line of symmetry between the parallel lines $l$ and $m$. Assume that $n$ was constructed at $E$ as on page 409 of the text. This means that there exists $D$ on $l$ and $F$ and $m$ such that both $\sphericalangle ED\Omega$ and $\sphericalangle EF\Omega$ are right and $\sphericalangle DE\Omega \cong \sphericalangle FE\Omega$ for some point $E$ between $l$ and $m$. Hence, $\triangle DE\Omega \cong \triangle FE\Omega$ by AA. Thus, $\overline{DE} \cong \overline{FE}$.

Now let $A$ be on $l$ and find $B$ on $n$ such that $\overline{AB}$ is perpendicular to $l$. Next find $C$ on $m$ such that $\overline{BC}$ is perpendicular to $m$. Join $\overline{DB}$ and $\overline{FB}$. Therefore, $\triangle EDB \cong \triangle EFB$ by SAS. Then, $\overline{BD} \cong \overline{FB}$ and $\sphericalangle BDA \cong \sphericalangle BFC$, so $\triangle BDA \cong \triangle BFC$ by AAS. We conclude that $\overline{AB} \cong \overline{BC}$.

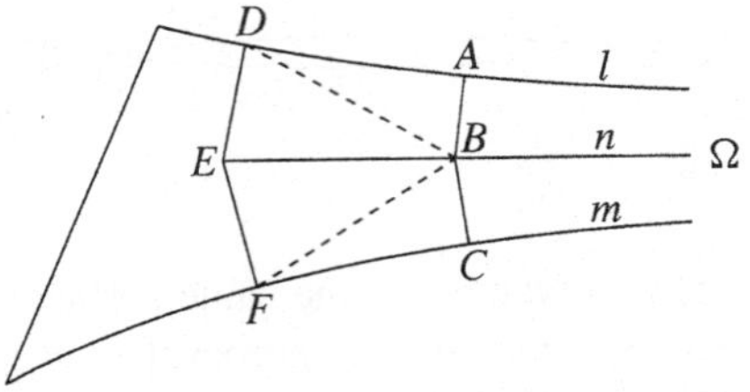


**15** Let $\mathcal{L}$ and $\mathcal{L}'$ be concentric horocycles with $\mathcal{L}$ being outer. Let $A$ and $B$ be on $\mathcal{L}$ and $A'$ and $B'$ be on $\mathcal{L}'$ such that $\overline{AA'}$ and $\overline{BB'}$ are common axes. By Theorem 12.2.9,

$$\mathrm{m}\widehat{A'B'} = \mathrm{m}\widehat{AB} \cdot e^{-k(AA')},$$

for some $k > 0$. Thus, $\mathrm{m}\widehat{A'B'} < \mathrm{m}\widehat{AB}$. Since arcs along a horocycle are measured using the intercepted segments, $A'B' < AB$.

**17** Let $O$ be the center of three concentric circles. Let $A$ and $B$ be on the outer circle, $A'$ and $B'$ on the middle circle, and $A''$ and $B''$ on the inner circle. Let

$$\begin{aligned} \theta &= \mathrm{m}\sphericalangle AOB, \\ r &= OA'', \\ s &= A''A', \\ t &= A'A. \end{aligned}$$

Then, by Equation 12.5,

$$\begin{aligned} \mathrm{m}\widehat{AB} &= (r+s+t)\theta, \\ \mathrm{m}\widehat{A'B'} &= (r+s)\theta, \\ \mathrm{m}\widehat{A''B''} &= r\theta. \end{aligned}$$

Suppose that

$$\frac{\mathrm{m}\widehat{AB}}{\mathrm{m}\widehat{A'B'}} = \frac{\mathrm{m}\widehat{A'B'}}{\mathrm{m}\widehat{A''B''}}.$$

That is,

$$\frac{(r+s+t)\theta}{(r+s)\theta} = \frac{(r+s)\theta}{r\theta}.$$

This gives

$$r^2 + rs + rt = r^2 + 2rs + s^2.$$

Hence,

$$t = \frac{s(r+s)}{r}.$$

Only under this condition does the equality hold. Otherwise, the proportion is false.

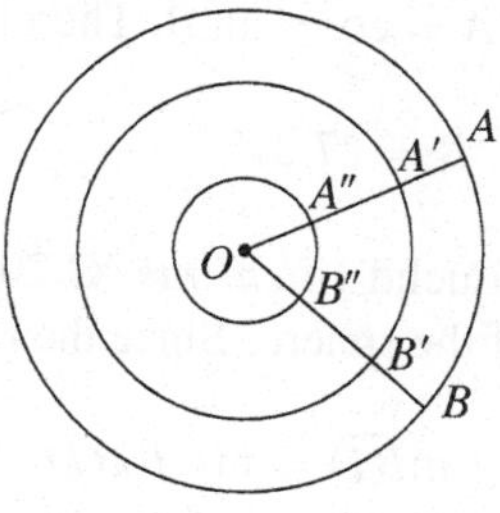


**19(a)** It has two branches because there is the set of points at a fixed distance from the line on one side of the line and there is a set of points that is equidistant to the line but is on the opposite side from the original set.

**21** Let $l$ be a line and $\mathscr{C}$ an equidistant curve to $l$. Let $\overline{AB}$ be an axis with $A$ on $\mathscr{C}$ and $B$ on $l$. Let $C$ and $D$ be on $\mathscr{C}$ so that $\overline{CA} \cong \overline{AD}$. Find $C'$ and $D'$ on $l$ such that $\overline{CC'}$ and $\overline{DD'}$ are perpendicular to $l$. Therefore, $\measuredangle CC'B \cong \measuredangle DD'B$. Since $\mathscr{C}$ is an equidistant curve to $l$, $\overline{CC'} \cong \overline{DD'}$. Therefore, $C'D'DC$ is a Saccheri quadrilateral. Hence, $\measuredangle C'CA \cong \measuredangle D'DA$ by Theorem 10.2.1.

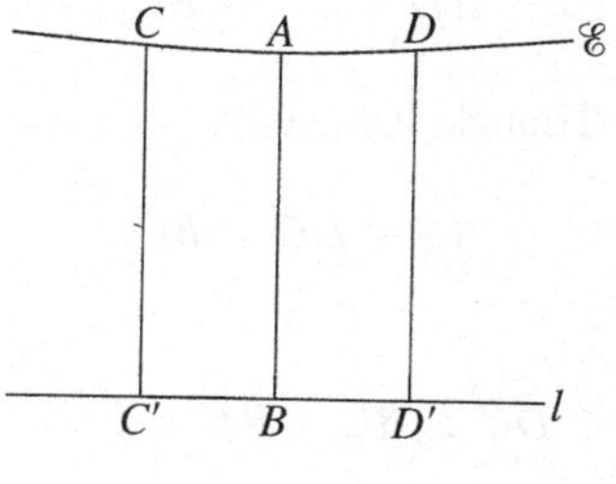


## 12.3 THE SURFACE OF A SPHERE

**1** Think of a globe. Suppose that two longitude lines (separated by, say, 10°) intersect the equator at $A$ and $B$ and intersect at the north pole $N$. Since $\widehat{AN}$ and $\widehat{BN}$ are perpendicular to the equator, $\triangle ABN$ has two right angles. If the longitude lines are separated by 90°, then $\triangle ABN$ has three right angles.

**3(a)** We know by Theorem 12.3.2 that

$$\cos c = \cos a \cos b$$

and

$$\sin b = \sin c \sin B.$$

Therefore, also by Theorem 12.3.2,

$$\cos A = \cos a \sin B = \frac{\cos c}{\cos b} \cdot \frac{\sin b}{\sin c} = \tan b \cot c.$$

**3(d)** By Theorem 12.3.2, $\cos A = \cos a \sin B$. Therefore, $\sec a = \sec A \sin B$.

**4(a)** $a = 67.73°, b = 25.50°, B = 27.27°$

**5(a)** This solution is due to Euclid (*Elements* XI.20). Let $\triangle BDC$ be on the unit sphere. Let $O$ be the center of the sphere. Since the radius of the sphere equals 1,

$$\begin{aligned} \mathrm{m}\widehat{BD} &= \mathrm{m}\sphericalangle BOD, \\ \mathrm{m}\widehat{DC} &= \mathrm{m}\sphericalangle DOC, \\ \mathrm{m}\widehat{BC} &= \mathrm{m}\sphericalangle BOC. \end{aligned}$$

If $\sphericalangle BOD \cong \sphericalangle DOC \cong \sphericalangle BOC$, any two sides of $\triangle BDC$ added together is twice the remaining side. So, suppose that $\widehat{BC}$ is the greatest; that is, suppose that $\sphericalangle BOC$ is the greatest. Find $E$ on $\overline{BC}$ such that $\sphericalangle BOE \cong \sphericalangle BOD$. We may assume that $\overline{OE} \cong \overline{OD}$ by extending $\overline{OB}$ and $\overline{OC}$ so that $B$ and $C$ are above the sphere while at the same time keeping $\sphericalangle DOB$, $\sphericalangle BOC$, and $\sphericalangle DOC$ fixed. Hence, $\triangle OBE \cong \triangle OBD$ by SAS. Thus, $\overline{BD} \cong \overline{BE}$. Since

$$BD + DC > BC$$

by the plane version of the Triangle Inequality,

$$BE + DC > BC.$$

Therefore,

$$DC > BC - BE = EC.$$

We conclude that $\mathrm{m}\sphericalangle DOC > \mathrm{m}\sphericalangle EOC$. Therefore,

$$\begin{aligned} \mathrm{m}\sphericalangle DOB + \mathrm{m}\sphericalangle DOC &> \mathrm{m}\sphericalangle DOB + \mathrm{m}\sphericalangle EOC \\ &= \mathrm{m}\sphericalangle BOE + \mathrm{m}\sphericalangle EOC \\ &= \mathrm{m}\sphericalangle BOC, \end{aligned}$$

which means that for $B$ and $C$ at their original positions on the surface of the sphere,

$$\mathrm{m}\widehat{BD} + \mathrm{m}\widehat{DC} > \mathrm{m}\widehat{BC}.$$

Similar work will obtain the other inequalities.

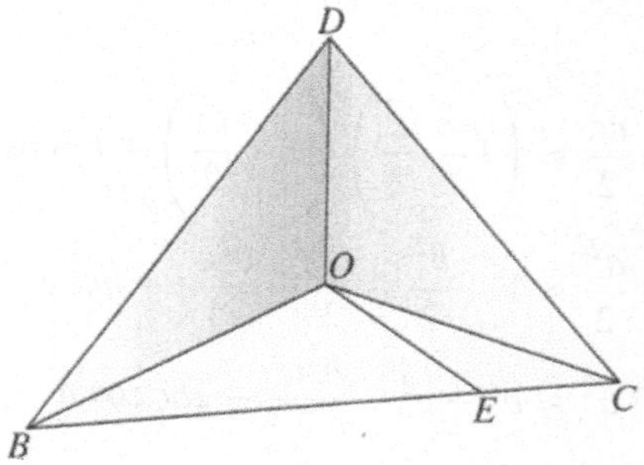


**5(c)** Using the diagram for the solution to Exercise 5(a), let $m\widehat{BD} = a$ and $m\widehat{BC} = b$. Suppose that

$$m\widehat{BD} < m\widehat{BC}.$$

This means that

$$m\measuredangle BOD < m\measuredangle BOC.$$

We can then find $E$ on $\overline{BC}$ such that $\measuredangle BOD \cong \measuredangle BOE$. Without loss of generality, assume that $\overline{OE} \cong \overline{OD}$. Hence, $\overline{BD} \cong \overline{BE}$ by SAS. We may then conclude that $BD < BC$, so

$$m\measuredangle BCD < m\measuredangle BDC.$$

This is what we want because $\measuredangle BCD = \measuredangle A$ and $\measuredangle BDC = \measuredangle B$ of $\triangle BCD$. To prove the converse, reverse the steps.

**5(e)** Because the measure of an angle is less than $\pi$, $A + B + C < 3\pi$. Also, we know that the area of a spherical triangle equals its angle excess. Therefore,

$$0 < \text{area}(\triangle ABC) = m\measuredangle A + m\measuredangle B + m\measuredangle C - \pi,$$

so

$$\pi < A + B + C.$$

**7** When we examine the triangle, we see that Equation 12.12 is equivalent to Exercise 3(a).

**9** Let $\triangle ABC$ be on the unit sphere. For $x$ small,

$$\cos x \approx 1 - \frac{x^2}{2} \text{ and } \sin x \approx x.$$

By Theorem 12.3.5,

$$\cos a = \cos b \cos c + \sin b \sin c \cos A.$$

Therefore,

$$1 - \frac{a^2}{2} \approx \left(1 - \frac{b^2}{2}\right)\left(1 - \frac{c^2}{2}\right) + bc\cos A,$$

$$1 - \frac{a^2}{2} \approx 1 - \frac{b^2}{2} - \frac{c^2}{2} + \frac{b^2c^2}{4} + bc\cos A,$$

$$a^2 \approx b^2 + c^2 - \frac{b^2c^2}{2} - 2bc\cos A,$$

$$a^2 \approx b^2 + c^2 - 2bc\cos A.$$

**12(a)** Let $\triangle A'B'C'$ be the **polar triangle** for $\triangle ABC$. This means that

$$a' = \pi - A,\ b' = \pi - B,\ c' = \pi - C, \text{ and } A' = \pi - a.$$

Therefore, by Theorem 12.3.5 we have

$$\cos a' = \cos b' \cos c' + \sin b' \sin c' \cos A',$$

$$\cos(\pi - A) = \cos(\pi - B)\cos(\pi - C) + \sin(\pi - B)\sin(\pi - C)\cos(\pi - a),$$

$$-\cos A = \cos B \cos C - \cos a \sin B \sin C.$$

From this we conclude that

$$\cos A = \cos a \sin B \sin C - \cos B \cos C.$$

## 12.4 HOROSPHERES

**1** This follows because the induced geometry on a horosphere is Euclidean.

**2(a)** Let $\mathcal{L}$ be a horocycle on the surface of a horosphere $\mathcal{F}$. Let $A$ be on $\mathcal{L}$, so $\overline{A\Omega}$ is an axis of $\mathcal{L}$ and $\mathcal{F}$. Let $B$ be on $\mathcal{F}$ but not on $\mathcal{L}$. Let $\Lambda$ be a plane that contains both $\overline{A\Omega}$ and $B$. The intersection of $\Lambda$ with $\mathcal{F}$ is a horocycle that contains both $A$ and $B$. Hence, $B$ corresponds with $A$.

**3** This is Bolyai's solution (*Appendix* 18). Let $A$, $B$, and $C$ be points on horosphere $\mathcal{F}$. Assume that all are not on the same horocycle. Let $\Omega$ be an ideal point so that $\overline{A\Omega}$, $\overline{B\Omega}$, and $\overline{C\Omega}$ are axes of $\mathcal{F}$. The planes $BA\Omega$ and $AC\Omega$ are different planes because $A$, $B$, and $C$ are not on the same horocycle. Let $M$ be the midpoint of $\overline{BA}$ and $N$ the midpoint of $\overline{AC}$. There exists an $F$ on $\mathcal{F}$ such that $\overline{FM}$ and $\overline{FN}$ are perpendicular to $\overline{BA}$ and $\overline{AC}$, respectively. Therefore, $\triangle FNA \cong \triangle FNC$ and $\triangle FMA \cong \triangle FMB$ by SAS, from which we conclude that

$$\overline{FB} \cong \overline{FA} \cong \overline{FC}.$$

Now find $H$ on the axis $\overline{F\Omega}$ such that $\overline{HA}$ is perpendicular to $\overline{F\Omega}$. Revolving $\overline{HA}$ about $\overline{F\Omega}$, keeping the $\measuredangle AHF$ constant, produces a circle that intersects $B$ and $C$.

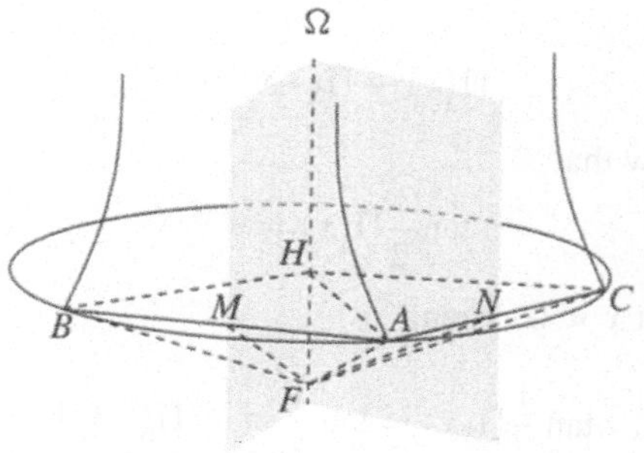


**5(b)** Let $\triangle ABC$ be as in the proof of Theorem 12.4.4 but such that $m\measuredangle A = \Pi(b')$, $m\measuredangle B = \Pi(c)$, $m\measuredangle C = \pi/2$, $BC = a$, $AC = \alpha$, and $AB = \beta$. We need to show that

$$\sin \Pi(b) = e^{kf(a)} \sin \Pi(c)$$

from the proof of Theorem 12.4.4. As in the proof, put in a horosphere so that we find

$$q = r \cos \Pi(b')$$

and

$$q = re^{kf(a)} \sin \Pi(c),$$

Hence,

$$r \cos \Pi(b') = re^{kf(a)} \sin \Pi(c).$$

By Exercise 5(a), we obtain the result.

**7** We see from the first diagram in the proof of Theorem 12.4.4 that $f(c) = BB''$, $f(a) = BD$, and $f(b) = CC'' = DB''$, where this last equation holds because the horocycles are concentric. Therefore,

$$f(c) = BB'' = BD + DB'' = f(a) + b(b).$$

**9** The first use of the word "similar" on page 433 is Exercise 5. The second use of the word "similar" refers to the proof of

$$e^{kf(a)} \sin \Pi(a) = e^{kf(b)} \sin \Pi(b).$$

## 12.5 EVALUATING THE PI FUNCTION

**1(b)** Since both $f(x) = e^{-kx}$ and $g(x) = \tan^{-1} x$ are continuous, $\Pi$ is continuous on its domain. Also, since

$$\frac{d}{dx}\Pi(x) = \frac{-2ke^{-kx}}{1 + e^{-2kx}}$$

is always negative, $\Pi$ is strictly decreasing.

**3** By definition,

$$\Pi(x) + \Pi(-x) = \pi.$$

Also, we already know that

$$\tan \frac{1}{2}\Pi(x) = e^{-kx}$$

holds for $x \geqslant 0$, so let $x < 0$. Then,

$$\begin{aligned}
\tan \frac{1}{2}\Pi(x) &= \tan \frac{1}{2}\left[\pi - \Pi(-x)\right] \\
&= \tan\left[\frac{\pi}{2} - \frac{1}{2}\Pi(-x)\right] \\
&= \cot \frac{1}{2}\Pi(-x) \\
&= \left[\tan \frac{1}{2}\Pi(-x)\right]^{-1} \\
&= e^{-kx}.
\end{aligned}$$

**5** As the area of the triangle increases and the vertices approach infinity, the angle measure of each vertex approaches 0. That is, the expression $\pi - A - B - C$ approaches $\pi$. Therefore, the area of the triangle approaches $\pi k^2$.

**7** By Exercise 5, the maximum area of a triangle (under **HA**) is $\pi k^2$. Since quadrilateral $ABCD$ is the disjoint union of $\triangle ABD$ and $\triangle BCD$ if convex, the maximum area of $ABCD$ is $2\pi k^2$. A more careful choice of triangles yields the same solution when $ABCD$ is not convex.

**9(a)** We already know that $\cos \Pi(x) = \tanh kx$, so generalizing to a complex domain with the appropriate Maclaurin series, we obtain

$$\begin{aligned}
\cos \Pi(xi) &= \tanh kxi \\
&= kxi - \frac{(kxi)^3}{3} + \frac{2(kxi)^5}{15} - \frac{17(kxi)^7}{315} + \cdots \\
&= i\left[kx + \frac{(kx)^3}{3} + \frac{2(kx)^5}{15} + \frac{17(kx)^7}{315} + \cdots\right] \\
&= i \tan kx.
\end{aligned}$$

**10(a)** By Equation 12.24,

$$\sin A \tan \Pi(a) = \sin B \tan \Pi(b),$$

but on a sphere of imaginary radius this equation becomes

$$\sin A \tan \Pi(ai) = \sin B \tan \Pi(bi).$$

However,

$$\tan \Pi(xi) = -i \csc kx,$$

so we obtain

$$\frac{\sin A}{-i \sin ka} = \frac{\sin B}{-i \sin kb}.$$

Hence,

$$\sin A \sin kb = \sin B \sin ka.$$

**11** We will check the formula for the circumference:

$$\begin{aligned}
\lim_{k\to 0} C &= \lim_{k\to 0} \frac{2\pi}{k} \sinh kr \\
&= \lim_{k\to 0} \frac{2\pi}{k} \left[ kr + \frac{k^3 r^3}{3} + \frac{k^5 r^5}{5} + \cdots \right] \\
&= \lim_{k\to 0} 2\pi \left[ r + \frac{k^2 r^3}{3} + \frac{k^4 r^5}{5} + \cdots \right] \\
&= 2\pi r.
\end{aligned}$$

# CHAPTER 13

# BERNHARD RIEMANN

## 13.1 METRIC SPACES

**1** We will finish the check that $\mathbb{R}$ is a metric space. On page 450 the Triangle Inequality was confirmed for $\mathbb{R}$. Here are the other three checks:

- $d[x, y] = |x - y| \geqslant 0$.
- $d[x, y] = 0$ if and only if $|x - y| = 0$ if and only if $x = y$.
- $d[x, y] = |x - y| = |y - x| = d[y, x]$.

**2** Let $x, y \in M$ such that $x \neq y$ and let $d$ be the metric on $M$. Define

$$\epsilon = \frac{d[x, y]}{2}.$$

In order to show that the intersection is empty, suppose that $a \in D(x, \epsilon) \cap D(y, \epsilon)$. Then,

$$d[a, x] < \frac{d[x, y]}{2}$$

and

$$d[a, y] < \frac{d[x, y]}{2}.$$

Therefore,

$$2d[x, y] \leqslant 2d[a, x] + 2d[a, y] < 2d[x, y],$$

a contradiction.

**5** Let $x, y, z \in M$. We will check the four conditions.

- Clearly, $d[x, y] \geqslant 0$.
- Also, $d[x, y] = 0$ if and only if $x = y$ by definition of $d$.
- If $x = y$, then $y = x$, so $d[x, y] = 0 = d[y, x]$. If $x \neq y$, then $y \neq x$, so $d[x, y] = 1 = d[y, x]$.
- First, suppose that $x = y$. Then $d[x, y] = 0$ which is less than or equal to $d[x, z] + d[y, z]$ since this value is either 0, 1, or 2. Next, assume that $x \neq y$. This means that $d[x, y] = 1$. Since either $x \neq z$ or $y \neq z$, the possible values of $d[x, z] + d[y, z]$ are 1 or 2.

**7(a)** $A$ is open in $\mathbb{R}$ because it is equal to the interval $(1, 2)$, which was shown to be open on page 452.

**7(c)** As a subset of $\mathbb{R}^2$, $A = \{(x, 0) : 1 < x < 2\}$. If we take $a \in A$ and any $\epsilon > 0$, $D(a, \epsilon) \cap A \neq \varnothing$ because the point $(a, \epsilon/2)$ is an element of $D(a, \epsilon)$ but not $A$.

**8(c)** Let $(x_0, y_0) \in \mathbb{R}^2$ but not on the line $2x - y - 1 = 0$. Using linear algebra techniques we can show that the distance from $(x_0, y_0)$ to the line is

$$\epsilon = \frac{|2x_0 - y_0 - 1|}{\sqrt{5}}.$$

Then,

$$D((x_0, y_0), \epsilon) \cap \{(x, y) \in \mathbb{R}^2 : y = 2x - 1\} = \varnothing,$$

so

$$D((x_0, y_0), \epsilon) \subseteq \mathbb{R}^2 \backslash \{(x, y) \in \mathbb{R}^2 : y = 2x - 1\}.$$

**9** Let $A = \{(x, y) \in \mathbb{R}^2 : a < x < b\}$. For any $x$ such that $a < x < b$, define $\epsilon = \min\{x - a, b - x\}$. To show that $D((x, y), \epsilon) \subseteq A$, take $(m, n) \in \mathbb{R}^2$ such that

$$(x - m)^2 + (y - n)^2 < \epsilon^2.$$

Then,

$$(x - m)^2 < \epsilon^2 - (y - n)^2 < \epsilon^2,$$

and so

$$|x - m| < \epsilon.$$

In other words,

$$-\epsilon < x - m < \epsilon.$$

We have two cases to check. First, if $\epsilon = x - a$, then $x - a \leqslant b - x$ which implies that $2x - a \leqslant b$. Therefore,

$$-x + a < x - m < x - a,$$

$$-2x + a < -m < -a,$$

and

$$a < m < 2x - a \leqslant b.$$

Hence, $(m, n) \in A$. The case when $\epsilon = b - x$ is demonstrated similarly.

**11(a)** Let

$$x \in \bigcup_{n=1}^{\infty} D(0, 1/n).$$

This means that there exists a positive integer $n_0$ such that $x \in D(0, 1/n_0)$. Since

$$D(0, 1/n_0) \subseteq \bigcup_{n=1}^{\infty} D(0, 1/n),$$

we see that the set is open.

**11(c)** $\bigcap_{n=1}^{\infty} D(0, 1/n) = \{0\}$, which is not an open set.

**13(b)** $\{(x, y) \in \mathbb{R}^2 : x^2 + y^2 < 1\}$

**15** Let $M$ be a metric space.

- To prove sufficiency, let $A$ be closed. We know by definition that $A \subseteq \text{cl}(A)$, so suppose that $a$ is not an element of $A$. By Theorem 13.1.12, $a$ is then not an accumulation point of $A$. Hence, there exists $\epsilon > 0$ such that
$$D(a, \epsilon) \cap A = \varnothing.$$
Since $D(a, \epsilon)^{\text{C}}$ is a closed superset of $A$, $a$ is not an element of $\text{cl}(A)$. We conclude that $\text{cl}(A) \subseteq A$, proving that $\text{cl}(A) = A$.

- For necessity, let $\text{cl}(A) = A$. Let $a$ be an accumulation point of $A$. Since $A \subseteq \text{cl}(A)$, $a$ is also an accumulation point of $\text{cl}(A)$. However, $\text{cl}(A)$ is closed, so $a \in \text{cl}(A)$. Hence, $a \in A$, so $A$ is closed.

**17(a)** The set is closed since its complement is $(-\infty, 0) \cup (0, \infty)$, which is open.

**17(d)** The complement of this set is $\{(x, y) \in \mathbb{R}^2 : x < 5 \text{ or } x > 7\}$, which is an open set.

**17(f)** The complement of the set is

$$S = \{(x, y, z) \in \mathbb{R}^3 : z \neq 1\}.$$

To show that $S$ is open, take $(a, b, c) \in S$. Let $\epsilon = |c - 1|$. We claim that $D((a, b, c), \epsilon)$ is a subset of $S$. Take $(x, y, z) \in \mathbb{R}^3$ such that

$$\sqrt{(a - x)^2 + (b - y)^2 + (c - z)^2} < \epsilon.$$

Hence, $|c - z| < |c - 1|$. We have four cases:

- If $c \geqslant z$ and $c > 1$, then $c - z < c - 1$, which yields $z > 1$.
- If $c < z$ and $c < 1$, then $z - c < 1 - c$, and we have $z < 1$.
- If $c < z$ and $c > 1$, then $1 < c < z$.
- If $c \geqslant z$ and $c < 1$, then $z \leqslant c < 1$.

In every case, $z \neq 1$, so $(x, y, z) \in S$.

**19** Fix $a \in M$ and let $C = \{b \in M : d[a, b] \leqslant 1\}$.

- Let $U = \{b \in M : d[a, b] > 1\}$. Because $U$ is the complement of $C$, showing that $U$ is open will prove that $C$ is closed. Let $x \in M$ such that $d[a, x] > 1$. Define $\epsilon = d[a, x] - 1$. Take $y \in d[x, \epsilon]$. Then,
$$d[x, y] < d[a, x] - 1 \leqslant d[a, y] + d[x, y] - 1,$$
which implies that
$$1 + d[x, y] < d[a, y] + d[x, y],$$
so $1 < d[a, y]$. Therefore, $y \in U$ and we have that $D(x, \epsilon) \subseteq U$. Hence, $U$ is open, and $C$ is closed.
- Let $x$ be an accumulation point of $C$. Then for all $\epsilon > 0$,
$$[D(x, \epsilon)\backslash\{x\}] \cap C \neq \varnothing,$$
so there exists in $M$ an element $w_\epsilon \neq x$ such that both $d[x, w_\epsilon] < \epsilon$ and $d[a, w_\epsilon] \leqslant 1$. Therefore, for all $\epsilon > 0$,
$$d[a, x] \leqslant d[x, w_\epsilon] + d[a, w_\epsilon] < \epsilon + 1.$$
Hence, $d[a, x] \leqslant 1$, and $x \in C$. This shows that $C$ is closed.

**21(a)** Let $x \in A$. Since the closure is the intersection of every closed superset of $A$, $x \in \mathrm{cl}(A)$.

**21(c)** Suppose that $x \in \mathrm{cl}(\varnothing)$. This means that $x$ is an element of every superset of $\varnothing$ that is closed. However, the empty set is closed and $\varnothing \subseteq \varnothing$. Thus, $x \in \varnothing$, a contradiction.

**21(e)** Suppose that $x \in \text{cl}(A)^{\text{C}}$. This means that there exists a closed set $F$ such that $A \subseteq F$ but $x \notin F$. Because $F^{\text{C}}$ is open, there exists $\epsilon > 0$ such that

$$D(x, \epsilon) \subseteq F^{\text{C}} \subseteq A^{\text{C}}.$$

Therefore, $x \in \text{int}(A^{\text{C}})$. The converse is essentially these steps reversed.

**23** Define $\mathcal{S} = \{f \in C[0, 1] : \|f\| < 1\}$. To show that $\mathcal{S}$ is an open subset of $C[0, 1]$, begin by letting $f \in C[0, 1]$ such that $\|f\| < 1$. Choose

$$\epsilon = \min\{\|f\|, 1 - \|f\|\}.$$

To prove that $D(f, \epsilon) \subseteq \mathcal{S}$, let $g \in D(f, \epsilon)$. First note that

$$d[f, g] = \int_0^1 |f(x) - g(x)|\, dx < \epsilon.$$

We have two cases to check. Suppose first that $\epsilon = \|f\|$. This means that

$$\|f\| \leqslant 1 - \|f\|,$$

and from this we get

$$\|f\| \leqslant \frac{1}{2}.$$

Therefore,

$$\begin{aligned} \|g\| = \int_0^1 |g(x)|\, dx &= \int_0^1 |g(x) - f(x) + f(x)|\, dx \\ &\leqslant \int_0^1 |f(x) - g(x)|\, dx + \int_0^1 |f(x)|\, dx \\ &< \|f\| + \|f\| \\ &\leqslant 1. \end{aligned}$$

If $\epsilon = 1 - \|f\|$, then

$$\int_0^1 |f(x) - g(x)|\, dx + \int_0^1 |f(x)|\, dx < (1 - \|f\|) + \|f\| = 1.$$

In either case, $g \in \mathcal{S}$.

## 13.2 TOPOLOGICAL SPACES

**1** Let $x \in f^{-1}[g^{-1}[V]]$. This means that there exists $y \in g^{-1}[V]$ such that $f(x) = y$ and there exists $z \in V$ such that $g(y) = z$. Hence, $x \in (g \circ f)^{-1}[V]$ because

$$(g \circ f)(x) = g(f(x)) = g(y) = z.$$

**3** $(X, \mathcal{T})$ is a topological space because $X, \varnothing \in \mathcal{T}$ and a systematic check of all possible unions and intersections of elements of $\mathcal{T}$ will yield only elements of $\mathcal{T}$.

**5** Let $\mathcal{T} = \{U : \mathbb{R}\backslash U \text{ is finite}\} \cup \{\varnothing\}$.

- $\mathbb{R} \in \mathcal{T}$ because $\mathbb{R}\backslash\mathbb{R} = \varnothing$ and $\varnothing \in \mathcal{T}$ by definition.
- Let $\mathcal{S} = \{U_i : i \in I\}$ be a subset of $\mathcal{T}$. If $\mathcal{S} = \{\varnothing\}$, then $\bigcup\mathcal{S} = \varnothing$, which is an element of $\mathcal{T}$. So, suppose that $\mathcal{S}$ contains a nonempty element. This means that there exists $i \in I$ such that $\mathbb{R}\backslash U_i$ is finite. Since $U_i \subseteq \bigcup\mathcal{S}$, the set $\mathbb{R}\backslash\bigcup\mathcal{S}$ is finite. Thus, $\bigcup\mathcal{S} \in \mathcal{T}$.
- Let $U_1, U_2, \ldots, U_k \in \mathcal{T}$. If one of the $U_i$ is empty, then $\bigcap_{i=1}^k U_i = \varnothing$, which means that this intersection is an element of $\mathcal{T}$. So, suppose that each $\mathbb{R}\backslash U_i$ is finite. Now,
$$\mathbb{R}\backslash\bigcap_{i=1}^{k} U_i \subseteq \bigcup_{i=1}^{k}(\mathbb{R}\backslash U_i).$$
Because
$$\left|\bigcup_{i=1}^{k}(\mathbb{R}\backslash U_i)\right| \leqslant |\mathbb{R}\backslash U_1| + \cdots + |\mathbb{R}\backslash U_k|,$$
$\mathbb{R}\backslash\bigcap_{i=1}^k U_i$ is finite because each cardinality on the right is finite. Therefore, the intersection is an element of $\mathcal{T}$.

**7** Let $f, g : \mathbb{R} \to \mathbb{R}$ be continuous. Take $\epsilon > 0$ and $x \in \mathbb{R}$. Because $f$ is continuous, there exists $\delta' > 0$ such that
$$\text{if } y \in D(g(x), \delta'), \text{ then } f(y) \in D(f(g(x)), \epsilon),$$
and because $g$ is continuous, there exists $\delta$ such that
$$\text{if } y \in D(x, \delta), \text{ then } g(y) \in D(g(x), \delta').$$
So, let $y \in D(x, \delta)$. Then,
$$g(y) \in D(g(x), \delta'),$$
and then
$$f(g(y)) \in D(f(g(x)), \epsilon).$$
We conclude that $f \circ g$ is continuous.

**9** Define the set
$$\mathcal{B} = \{(a, b) \times (c, d) : a < b, c < d\}.$$
Let $U$ be open in $\mathbb{R}^2$. For every $x \in U$, there exists $\epsilon_x > 0$ such that $D(x, \epsilon_x) \subseteq U$. Inside this disk there exists an open rectangle $R_x \in \mathcal{B}$ with the property that
$$x \in R_x \subseteq D(x, \epsilon_x).$$

Therefore,

$$U = \bigcup_{x \in U} R_x,$$

and we see that $\mathcal{B}$ is on open base for $\mathbb{R}^2$.

**11** The interval $[0, 2\pi)$ is a topological space with the relative topology

$$\mathcal{T} = \{A \cap [0, 2\pi) : A \text{ is open in } \mathbb{R}\},$$

where $\mathbb{R}$ is given the induced topology. Let $S^1$ denote the unit circle in $\mathbb{R}^2$. It has a topology consisting of the open arcs and unions of open arcs on the circle. Then,

$$f : [0, 2\pi) \to S^1$$

defined by $f(\theta) = (\cos\theta, \sin\theta)$ is continuous and one-to-one but does not preserve open sets. For example, $[0, \pi/2)$ is open in $[0, 2\pi)$ but $f[[0, \pi/2)]$ is not open on the unit circle. Therefore, $f$ is not a homeomorphism.

**13** Let $f$ be as described in the problem. We will show that $f$ is an open mapping. The proof that it is continuous will be similar. Let $V$ be an open subset of $X$. Since $\mathcal{B}$ is a base for $X$, there exists $J \subseteq I$ such that

$$V = \bigcup_{j \in J} U_j.$$

To prove that $f[V]$ is open, it suffices to show that

$$f[V] = \bigcup_{j \in J} U'_j.$$

Let $y \in f[V]$. This means there exists $x \in V$ such that $f(x) = y$. We know that $x \in U_{j_0}$ for some $j_0 \in J$. Hence, $y \in U'_{j_0}$, and then $y \in \bigcup_{j \in J} U'_j$. Since with a little modification, these steps reverse, we have that $f$ is an open map.

**15** The function

$$f : (0, 1) \to (5, 6)$$

defined by $f(x) = x + 5$ is a homeomorphism, and

$$g : (0, 1) \to \mathbb{R}$$

defined by

$$g(x) = \tan\left(\pi x - \frac{\pi}{2}\right)$$

is also a homeomorphism.

**17(a)** $X \backslash X = \varnothing$ and $X \backslash \varnothing = X$. Therefore, since both $X$ and $\varnothing$ are open, both $X$ and $\varnothing$ are closed.

**17(c)** Let $C_1, C_2, \ldots, C_k$ be closed. This means that $X \backslash C_i$ is open for each $i$. Therefore, since

$$X \backslash \bigcup_{i=1}^{k} C_i = \bigcap_{i=1}^{k} (X \backslash C_i)$$

and $\bigcap_{i=1}^{k} (X \backslash C_i)$ is open, $\bigcup_{i=1}^{k} C_i$ is closed.

**19** Let $A$ be a subset of the metric space $X$ with topology $\mathcal{T}$.

- To prove sufficiency, let $A$ be closed. We know by definition that $A \subseteq \mathrm{cl}(A)$, so take $a \in \mathrm{cl}(A)$. Since $\mathrm{cl}(A)$ is the intersection of all closed supersets of $A$ and $A$ is closed, $a \in A$. Hence, $\mathrm{cl}(A) \subseteq A$.
- For necessity, let $\mathrm{cl}(A) = A$. Since $\mathrm{cl}(A)$ is closed by Exercise 17(b), $A$ is closed.

## 13.3 STEREOGRAPHIC PROJECTION

**1** Assuming that $cr - d \neq 0$, the center of the circle is

$$\left( -\frac{r^2 a}{cr - d}, -\frac{r^2 b}{cr - d} \right),$$

and its radius is

$$\sqrt{\frac{r^2(a^2r^2 + b^2r^2 + c^2r^2 - d^2)}{cr - d}}.$$

**2(a)** The plane $y = 0$ intersects the sphere of radius $r = 1$ yielding the given circle. Therefore, in Equation 13.14, $a = 0$, $b = 1$, $c = 0$, and $d = 0$. We see that $cr - d = 0$, so the image is a line. Its equation is $v = 0$. This is the $x$-axis, as we should have expected if we had thought through the geometry.

**2(e)** This circle is formed when the sphere is cut by the plane:

$$\frac{2}{9}y + \frac{8 - 6\sqrt{3}}{9}z = \frac{4 - 2\sqrt{3}}{9}.$$

**3(a)** To find the plane that contains $\widehat{AB}$, we need three points. Since we are looking for the great circle that contains $A$ and $B$, the third point will be the origin. So, $\langle 4/3, 0, 0 \rangle$ and $\langle 2/3, 2/3, 1/3 \rangle$ are two vectors in the desired plane. Since,

$$\mathbf{m} = \left\langle 0, \frac{8}{9}, -\frac{16}{9} \right\rangle = \left\langle \frac{4}{3}, 0, 0 \right\rangle \times \left\langle \frac{2}{3}, \frac{2}{3}, \frac{1}{3} \right\rangle,$$

the equation for the plane is

$$\frac{8}{9}y - \frac{16}{9}z = 0.$$

Likewise, the equation for the plane that forms the great circle containing $\widehat{AC}$ is

$$\frac{2-\sqrt{3}}{3}x - \frac{2}{3}y + \frac{2\sqrt{3}}{3}z = 0$$

since its normal vector is

$$\mathbf{n} = \left\langle \frac{2-\sqrt{3}}{3}, -\frac{2}{3}, \frac{2\sqrt{3}}{3} \right\rangle.$$

**3(c)** If we let $\theta$ represent the angle between the normal vectors to the planes that cut the great circles that form the side of the angle, using the notation from the previous solution we have

$$\theta = \cos^{-1}\left(\frac{\mathbf{m} \bullet \mathbf{n}}{\|\mathbf{m}\| \cdot \|\mathbf{n}\|}\right).$$

Therefore, the measure between the great circles is either 174.87° and 5.13° (Figure 13.7), which means that $\theta = 5.13°$.

**3(d)** To simply our notation, we will view $\mathbb{R}^2 \times \{0\}$ as $\mathbb{R}^2$. The images of the two great circles in $\mathbb{R}^2$ are given by the equations

$$x^2 + y^2 - y - 1 = 0$$

and

$$x^2 + y^2 + \frac{2-\sqrt{3}}{\sqrt{3}}x - \frac{2}{\sqrt{3}}y - 1 = 0.$$

The centers of these circles are $C(0, 1/2)$ and $D([-2+\sqrt{3}]/2\sqrt{3}, 1/\sqrt{3})$, respectively. Since $E(1, 1)$ is a point of intersection of the two circles, we need to find the measure of the angle between $\overline{EC}$ and $\overline{ED}$. We do this using the vectors

$$\mathbf{u} = \left\langle 1, \frac{1}{2} \right\rangle$$

and

$$\mathbf{v} = \left\langle \frac{\sqrt{3}+2}{2\sqrt{3}}, \frac{\sqrt{3}-1}{\sqrt{3}} \right\rangle.$$

Since

$$\cos\theta = \frac{\mathbf{u} \bullet \mathbf{v}}{\|\mathbf{u}\| \cdot \|\mathbf{v}\|},$$

the angle between the circles is either 174.87° or 5.13°.

**5** If $r$ is the radius of the sphere, the image of the upper hemisphere is

$$\{(x, y, 0) : x^2 + y^2 \geqslant r^2\},$$

and the image of the front hemisphere is

$$\{(x, y, 0) : x \geqslant 0\}.$$

**7** As in Figure 13.7, we will measure the angle formed by the two circles using $\sphericalangle ADB$. In the figure below,

$$A \text{ is } (0,\ 2),\ B \text{ is } (2,\ -2),\ AD = 3,\ BD = 4,$$

and

$$AB = 2\sqrt{5} \approx 4.4721.$$

From the figure we have the system

$$EB^2 + ED^2 = 16,$$
$$AE^2 + ED^2 = 9.$$

Therefore,

$$EB^2 - AE^2 = 7.$$

Since $AE = AB - EB$,

$$EB^2 - EB^2 + 2EB \cdot AB - AB^2 = 7,$$

from which follows

$$EB = \frac{27}{4\sqrt{5}} \approx 3.0187.$$

Lastly, we compute that

$$AE \approx 1.4534 \text{ and } ED \approx 2.6244.$$

Now, the line joining the centers of the circles in parametric form is

$$x = 2 - 2t,$$
$$y = -2 + 4t.$$

Since $EB/AB \approx .6750$, the point $E$ has coordinates $(.6500,\ .7000)$. The line through $E$ that is perpendicular to $\overline{AB}$ has the equation

$$x = .65 + 4t,$$
$$y = .7 + 2t.$$

This line intersects $\odot A$ when $t = \pm .5869$. That is, the point $D$ has coordinates $(2.9976,\ 1.8738)$.

We are now almost finished. $\langle -2.9976,\ 1.1262 \rangle$ points from $D$ to $A$ and has a length of 3.0003. $\langle -.9976,\ -3.8738 \rangle$ points from $D$ to $B$ and has a length of 4.0002. Thus,

$$\cos(\sphericalangle ADB) = \frac{\langle -2.9976,\ .1262 \rangle \bullet \langle -.9976,\ -3.8738 \rangle}{3.0003 \cdot 4.0002} = .2084,$$

which tells us that $\theta \approx 77.9714°$ (about $.4332\pi$ radians).

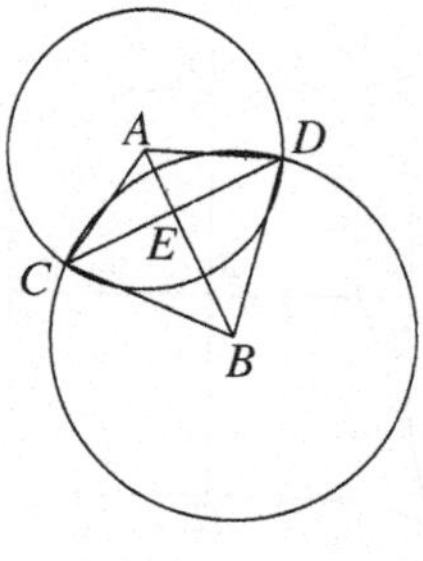


## 13.4 CONSISTENCY OF NON-EUCLIDEAN GEOMETRY

**1** The point at infinity is the ideal point $\Omega$ on the asymptote. The geodesic between any two points on the surface of the pseudosphere serves as a line in the model. Longitude lines, such as $\overline{A\Omega}$ and $\overline{B\Omega}$, are the parallel lines. The line $\overline{CD}$ is an example of an intersecting line relative to the parallels $\overline{A\Omega}$ and $\overline{B\Omega}$.

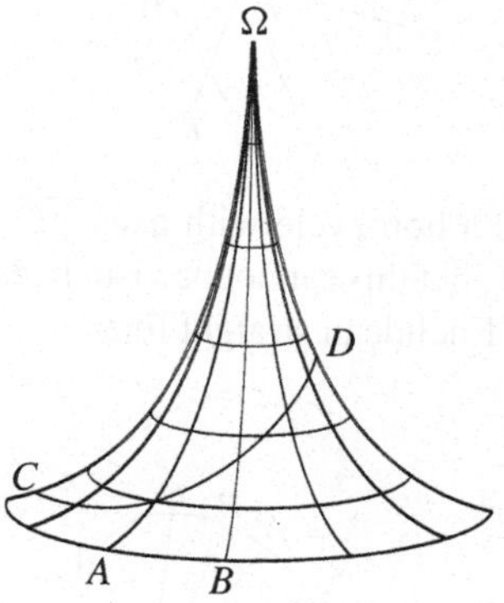


**2(b)** The significance of 12.56 is the it is approximately equal to $4\pi$.

**3** Construct a Lambert quadrilateral $ABCD$ such that $\overline{AB} \cong \overline{AD}$. Join $\overline{DB}$. Since $\triangle ABD$ is isosceles, $\measuredangle ADB \cong \measuredangle ABD$, and since $\measuredangle ADC$ and $\measuredangle ABC$ are both right, $\measuredangle BDC \cong \measuredangle DBC$ by angle subtraction. Therefore, $\triangle CDB$ is isosceles with base $\overline{DB}$, so we may conclude that $\overline{DC} \cong \overline{CB}$. Now construct three copies of $ABCD$: $AFED$, $AFGH$, and $ABIH$. We see that

$$\measuredangle GEC \cong \measuredangle ECI \cong \measuredangle CIG \cong \measuredangle IGE$$

and

$$\overline{GE} \cong \overline{EC} \cong \overline{CI} \cong \overline{GI}.$$

Therefore, $ECIG$ is a pseudosquare.

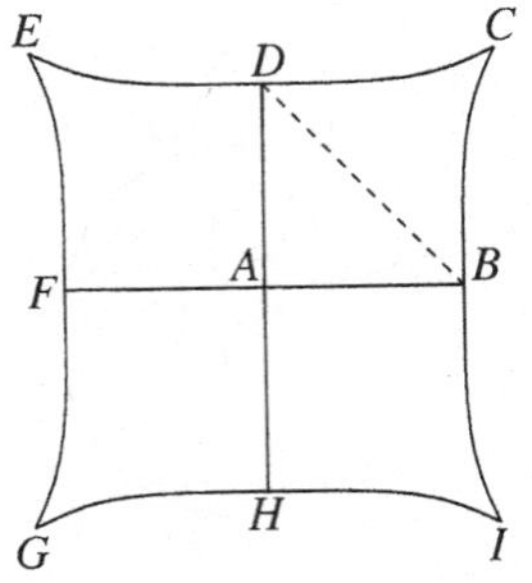


**5** In this Klein model, $\overline{AP}$ and $\overline{BP}$ are parallel to $l$, $\overline{CD}$ is a nonintersecting line of $l$, $\overline{EF}$ intersects $l$, $Q$ is a point, and $PGH$ is a triangle.

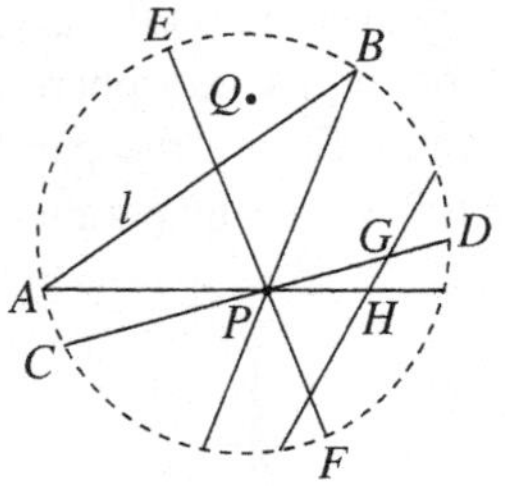


**7** In this Klein model, $\mathscr{L}$ is a horocycle with axes $\overline{A\Omega}$, $\overline{B\Omega}$, and $\overline{C\Omega}$. These axes should be perpendicular to $\mathscr{L}$, but this cannot be visualized in the Klein model because the lines are represented by Euclidean straight lines.

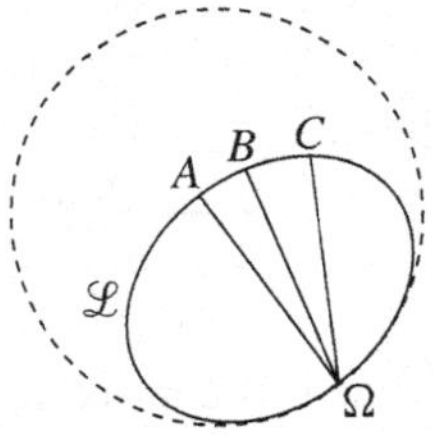


**9** It is not Euclidean because given a line $\overline{AB}$ and a point $P$ off the line, there is more than one line through $P$ that does not intersect $\overline{AB}$. The parallel to $\overline{AB}$ (on the right) is $\overline{PB}$. However, it appears, that the angle sum of a triangle in this model equals two right angles. This cannot serve as a model of **HR** or **HA** unless a metric is defined for angle measure so that the measure of the angles of the triangle becomes less than two right angles.

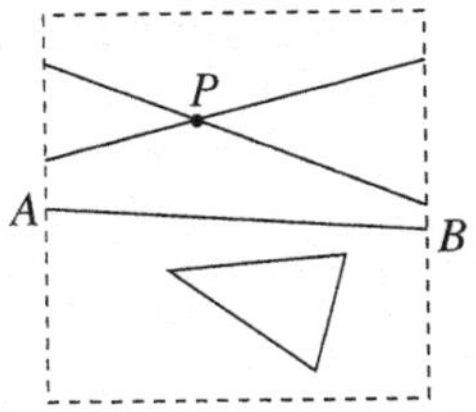

**11** From the proof of Theorem 13.4.2 we arrive at the concept of an absolute pole. Let $l$ be a line. Let $N$ be the absolute pole on one "side" of $l$ and $S$ the absolute pole on the other "side" of $l$. In single elliptic geometry, $N$ and $S$ are actually the same point. So if we draw a line from $l$ to join with $N$, we find that the line will continue from $S = N$ from the other "side," looping back and reconnecting with $l$. Hence, $l$ cannot separate the plane into two half planes. A flat model of this behavior is the Möbius strip.

**13** It can.

**15** It all depends upon the geometry. In elliptic geometry, there is no maximum measure for the third angle, but the total angle sum cannot exceed $3\pi$. In Euclidean geometry, if we allow ideal triangles, the third angle has measure zero, and two of the sides are parallel. In hyperbolic geometry, there can be no triangles with two right angles.

# CHAPTER 14

# JEAN-VICTOR PONCELET

## 14.1 THE PROJECTIVE PLANE

**1(a)** Use the Fano plane.

**1(c)** The Fano plane shows that it is possible for all four postulates to hold.

- Postulate 1 can be false but the others true



- Postulate 2 can be false but the others true

- Postulate 3 can be false but the others true

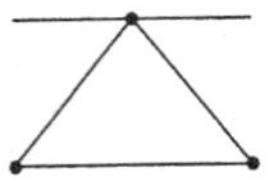

- Postulate 4 can be false but the others true

**3(a)** Let $l$ and $m$ be two distinct lines. By Postulate 3, there exists a point $P$ on both $l$ and $m$. Suppose that there exists a point $Q$ different from $P$ so that $Q$ is also on $l$ and $m$. This means that the pair $P$ and $Q$ are on two distinct lines, contradicting Postulate 2.

**3(c)** Suppose that there is a line $l$ with three distinct points on it. By Postulate 4, we know that there is a point not on $l$, so that we have at least four points in contradiction of Postulate 1. Therefore, no line can have three distinct points.

**3(e)** By Postulate 1 and Exercise 3(b), there are exactly three points, call them $A$, $B$, and $C$, and there are exactly three lines, call them $l$, $m$, and $n$. Assume that these elements are labeled in such a way that $A$ and $B$ are on $l$, $A$ and $C$ are on $m$, and $B$ and $C$ are on $n$. With this choice we see that $\{A, B\}$ is the desired set.

**5(c)** It is a projective geometry. Postulate 4 implies Postulate 14.1.1.1, Postulate 5 implies Postulate 14.1.1.2, Postulate 2 implies Postulate 14.1.1.3, and Exercise 6(b) implies Postulate 14.1.1.4

**6(b)** We first show that there are at least seven points and at least seven lines.

- By Postulates 1 and 2, there exists a line with three points on it, $E$, $F$, and $G$.
- By Postulate 3 there is a point $A$ not on $\overline{EFG}$.
- By Postulates 4 and 2, there exists $\overline{AE}$ with $B$ on it, $\overline{AF}$ with $C$ on it, and $\overline{AG}$ with $D$ on it. We now have seven points and four lines.
- By Postulate 4, join $\overline{ED}$ and $\overline{BG}$. We now have six lines. They must intersect $\overline{AF}$ (Postulate 5), and they both must intersect $\overline{AF}$ in $C$, else either Postulate 2 or Postulate 4 is contradicted.
- By Postulate 4, join $\overline{BD}$ and observe that $F$ must be on this line by the argument in the previous step. We now have seven lines, and Fano's geometry at least contains a copy of the Fano plane.

We now show that there are at most seven points and at most seven lines. Suppose that there is another point $P$ in the geometry besides those constructed above. It cannot lie on any of the already existing lines because that would contradict Postulate 2. Join $\overline{AP}$ by Postulate 4. This new line must intersect $\overline{EG}$ at $F$ (Postulates 2 and

4). However, the new must also intersect $\overline{ED}$, which means that the new line has at least four points on it, contradicting Postulate 2. Therefore, there cannot be an eighth point. This argument also shows that there cannot be an eighth line.

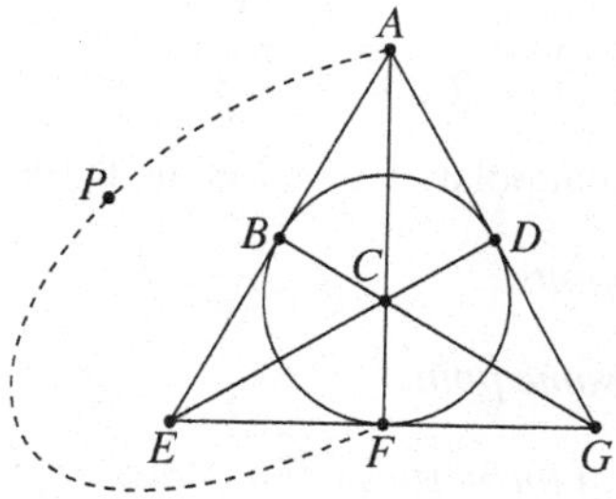


**7(a)** Young's postulates are satisfied in the following model. There are nine points and 12 lines. The lines are $\overline{ABC}$, $\overline{FED}$, $\overline{GHI}$, $\overline{AFG}$, $\overline{BEH}$, $\overline{CDI}$, $\overline{AEI}$, $\overline{CEG}$, $\overline{BFI}$, $\overline{FHC}$, $\overline{HDA}$, $\overline{BDG}$.

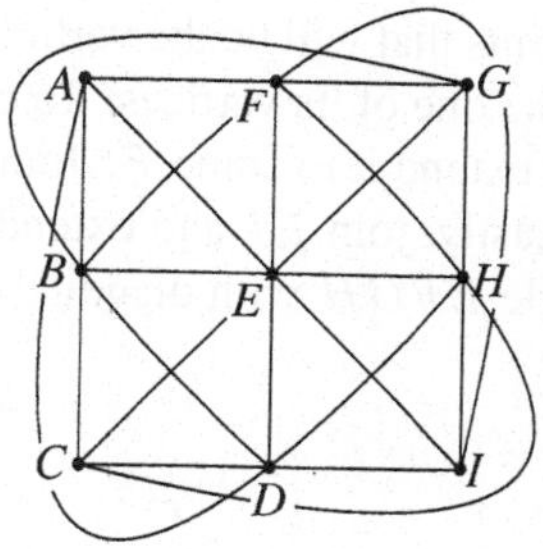


**7(c)** Young's geometry is not projective because it has parallel lines.

**9** This follows by Postulates 14.1.1.1 and 14.1.1.2.

**11(a)** Let $\phi : \mathcal{P} \to \mathcal{P}'$ be a collineation. Take $P', Q' \in \mathcal{P}'$. Since $\phi$ is a one-to-one correspondence, find the unique points $P, Q \in \mathcal{P}$ such that $\phi(P) = P'$ and $\phi(Q) = Q'$. In other words, $\phi^{-1}(P') = P$ and $\phi^{-1}(Q') = Q$.

- Pick $R'$ on $\overline{P'Q'}$ and find $R \in \mathcal{P}$ such that $\phi(R) = R'$. We may translate this as $\phi(R) \in \overline{\phi(P)\phi(Q)}$. Since $\phi$ is a collineation, $R \in \overline{PQ}$. That is, $\phi^{-1}(R) \in \overline{\phi^{-1}(P')\phi^{-1}(Q')}$ because $\phi^{-1}(R') = R$.

- Now let $\phi^{-1}(R') \in \overline{\phi^{-1}(P')\phi^{-1}(Q')}$, so $R \in \overline{PQ}$. Since $\phi$ is a collineation, $\phi(R) \in \overline{\phi(P)\phi(Q)}$, which means that $R' \in \overline{P'Q'}$.

We conclude that $\phi^{-1}$ is a collineation.

**12** Only incidence geometry has at least two nonisomorphic models.

## 14.2 DUALITY

**1(b)** One the tangents to the ellipse at $x = 2$ has the equation

$$y = \frac{-1}{2\sqrt{3}}x + \frac{4}{\sqrt{3}}.$$

Therefore, the pole of the vertical polar at $x = 2$ is the is $(8, 0)$.

**2(a)** *There exist at least two points.*

**2(c)** *Not all lines contain the same point.*

**2(e)** *The point joins with one of the vertices of a complete quadrilateral.*

**2(g)** *All lines contain either the first or second point.*

**3** Given a hexagon circumscribed about a circle, the lines joining opposite vertices are concurrent.

**5** Let $A$, $B$, and $I$ be fixed points that will be the vertices of the diagonal triangle of a complete quadrangle. Let $E$ be one of its vertices. To construct the quadrangle, start by joining $\overline{EB}$. Join $\overline{AE}$ and extend it to some $F$. Join $\overline{FI}$ and extend it to intersect $\overline{EB}$ at some $H$. Join $\overline{BF}$. Lastly, join $\overline{EI}$ and extend it to intersect $\overline{FB}$ at $G$. We now have complete quadrangle $EFGH$ with diagonal $\triangle ABI$.

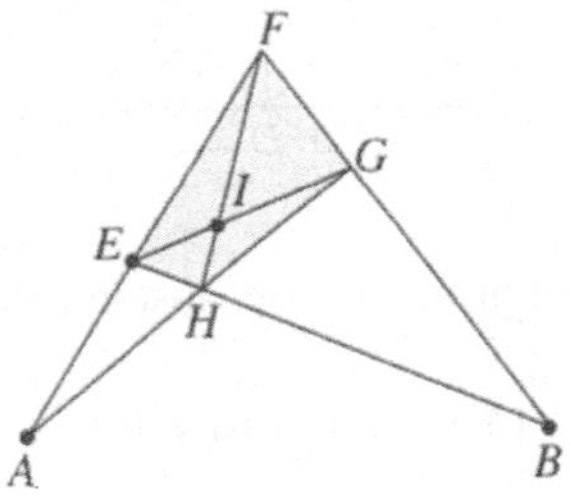


**7** Assume $\mathsf{H}(AB, CD)$. This means that $\mathsf{Q}(ABC, ABD)$. This implies $(AA)(BB)(CD)$, which is equivalent to $(AA)(CD)(BB)$. Hence,

$$\frac{AC \cdot AC}{AD \cdot AD} = \frac{BC \cdot BC}{BD \cdot BD}.$$

That is,

$$\frac{AC \cdot BD}{AD \cdot BC} = 1.$$

Reverse the steps to prove the converse.

**9** Suppose that $\mathsf{H}(AB, CD)$ and $C$ is the midpoint of $\overline{AB}$. This means that $AB = 2AC$. Therefore, by Exercise 7,

$$\frac{AC \cdot BD}{2AC \cdot AD} = \frac{BD}{2AD} = 1,$$

but this implies that $BD = 2AD$. Since $C$ is the midpoint of $\overline{AB}$, the only way that this is possible is if $D$ is an ideal point.

**11** View $O$ as an origin and note that

$$(OB - OA)(OB' - OC)(OA' - OC')$$

equals

$$(OB' - OA')(OB - OC')(OC - OA).$$

Therefore,

$$\begin{aligned}(AB')(BC) &= \frac{AB \cdot B'C}{AC \cdot B'B} \\ &= \frac{(OB - OA)(OB' - OC)}{(OC - OA)(OB' - OB)} \\ &= \frac{(OB' - OA')(OB - OC')}{(OA' - OC')(OB' - OB)} \\ &= \frac{(OB' - OA')(OB - OC')}{(OC' - OA')(OB - OB')} \\ &= \frac{A'B' \cdot BC'}{AC' \cdot BB'} \\ &= (A'B)(B'C').\end{aligned}$$

Hence, $\mathsf{H}(AB', BC)$ if and only if $\mathsf{H}(A'B, B'C')$.

**13(a)** (24/13, 0)

**15** Either take inspiration from the method of finding the harmonic conjugate of three collinear points or use Exercise 17. To use the second method, let $a$, $b$, and $c$ be concurrent lines through a point $P$. Let $l$ intersect the three lines but not contain $P$. Let the points of intersection be $A$, $B$, and $C$. Let $D$ be on $l$ such that $\mathsf{H}(AB, CD)$. Then $a$, $b$, $c$, and $\overline{DP}$ form a harmonic set of lines.

**17** Let $a$, $b$, $c$, and $d$ be concurrent at $P$ and intersect a line $l$ at the points $A$, $B$, $C$, and $D$, respectively. By the Principle of Duality we only need to show that $\mathsf{H}(AB, CD)$ implies $\mathsf{H}(ab, cd)$, so assume that $A$, $B$, $C$, and $D$ form a harmonic set of points on $l$. We can then construct a complete quadrangle $PQRS$ as illustrated in the diagram so that $D$ is on $\overline{PR}$. Observe that we also have a complete quadrilateral with vertices $C$, $B$, $R$, $Q$, $S$, and $A$ with a diagonal trilateral consisting of $\overline{AS}$, $\overline{BQ}$, and $\overline{CR}$. Further notice that $a$ and $b$ are the first two of these lines while $C$ is on $c$ and $R$ is on $d$. Hence, $a$, $b$, $c$, and $d$ form a harmonic set of lines.

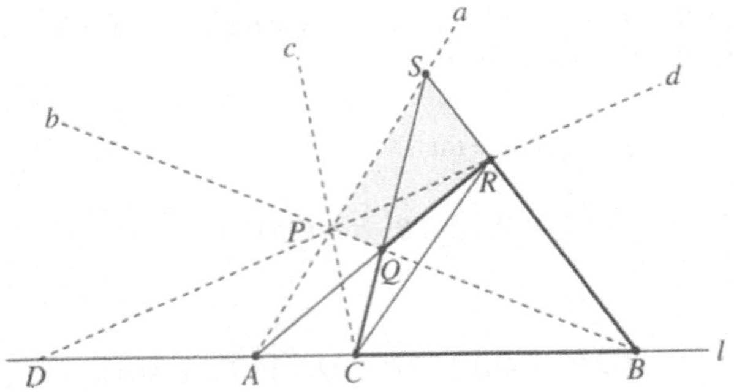


## 14.3 PERSPECTIVITY

**1** Let $T : l \to m$ be a central perspectivity with center $O$. Since $l$ and $m$ are different, they intersect in exactly one point (Theorem 14.2.1). Since these two lines each have at least three points (Postulate 14.1.1.3), there exist $A$ and $B$ on $l$ but not on $m$ and $A'$ and $B'$ on $m$ but not on $l$. Suppose that $T(A) = A'$ and $T(B) = B'$. We may assume this because if $C$ is on both $l$ and $m$, $T(C) = C$. The center of perspectivity for $T$ is on both $\overline{AA'}$ and $\overline{BB'}$. This means that the unique intersection of these two lines is the center.

**2** This is false. Every two distinct lines share exactly one point. If two lines share more than one point, the lines are the same.

**3** Let $T : l \to l$ be a perspectivity with center $O$. By definition, $O$ is not on $l$. Let $A$ be on $l$ and join $\overline{AO}$. This line intersects $l$ only once (Theorem 14.2.1). Therefore, $T(A) = A$. Since $A$ was arbitrarily chosen, $T$ is the identity map.

**5** Suppose that $T : l \to m$ is a central perspectivity. Let $A$ and $B$ be distinct points on $l$ and $A'$ and $B'$ distinct points on $m$. Join $\overline{AA'}$ and $\overline{BB'}$. The center $O$ of the perspectivity $T$ is the unique intersection of $\overline{AA'}$ and $\overline{BB'}$. Let $C$ be any point on $l$. Since $\overline{OC}$ intersects $m$ in a unique point $C'$, we conclude that $T(C) = C'$. Hence, $T$ is determined uniquely by the two points $A$ and $B$.

**6(a)** The dual of the center of a central perspectivity is the **axis** of an axial perspectivity. Use this new term in the definition.

**6(b)** Let $\mathcal{L}$ be a set of lines concurrent at $P$ and $\mathcal{M}$ a set of lines concurrent at $Q$. Let $T : \mathcal{L} \to \mathcal{M}$. Let $o$ be the axis of $T$.

- To show that $T$ is one-to-one, let $l,\ m \in \mathcal{L}$. Write $l' = T(l)$ and $m' = T(m)$. Suppose that $l' = m'$. Let $R$ be the point of intersection of $l'$ with $o$. The lines $l$ and $m$ cannot be different because this would mean that the points $P$ and $R$ are on two distinct lines. Hence, $l = m$.

- To demonstrate that $T$ is onto, take $l' \in \mathcal{M}$. Let $R$ be the intersection of $l'$ with $o$. Join $\overline{PR}$. Then, $T(\overline{PR}) = l'$.

Thus, $T$ is a one-to-one correspondence.

**7** A central perspectivity is a function between two ranges such that corresponding points lie on the same line in a pencil of lines.

**9** Since a projectivity is uniquely determined by three points, we need to focus on the images of three points. We consider two cases (are there others?): First suppose that $T(A) = B, T(B) = A$, and $T(C)$. The three perspectivities with centers $O$, $O'$, and $O''$ as in the diagram show that $T$ is the composition of three central perspectivities.

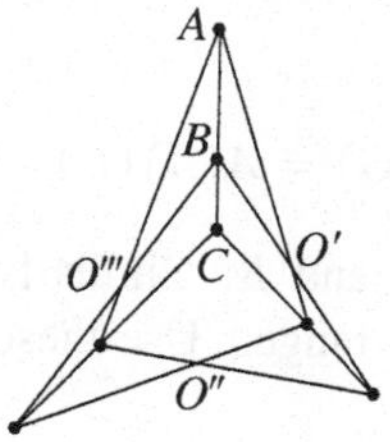


Now let $T(A) = B, T(C) = B$, and $T(C) = A$. Choose two new points and call them $E$ and $F$. Define perspectivity

$$T_1 : \overline{ABC} \rightarrow \overline{EDC}$$

with center $O'$ by

$$T_1(A) = E,\ T_1(B) = D,\ T_1(C) = C.$$

Define

$$T_2 : \overline{EDC} \rightarrow \overline{BDF}$$

with center $O''$ by

$$T_2(E) = B,\ T_2(D) = D,\ T_2(C) = F.$$

Lastly, define

$$T_3 : \overline{BDF} \rightarrow \overline{BCA}$$

with center $O'''$ by

$$T_3(B) = B,\ T_3(D) = D,\ T_3(F) = A.$$

With these choices, $T = T_3 \circ T_2 \circ T_1$.

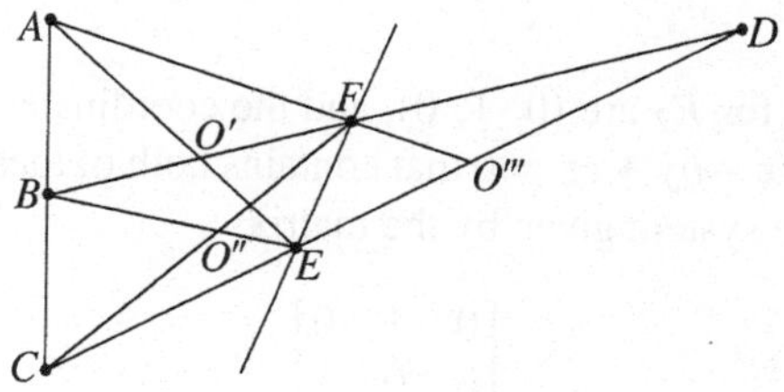

**11** In the familiar diagram below, we have

$$T_1 : \overline{DACB} \to \overline{DEFG}$$

with center $O'$ defined by

$$T_1(A) = E,\ T_1(B) = G,\ T_1(C) = F,\ T_1(D) = D.$$

In addition, we have

$$T_2 : \overline{DEFG} \to \overline{DACB}$$

with center $O''$ defined by

$$T_2(E) = B,\ T_2(G) = A,\ T_1(F) = C,\ T_1(D) = D.$$

Therefore, $T_2 \circ T_1$ exchanges $A$ and $B$. Similarly projectivities can be defined to exchange any two points of the range. Use these functions to obtain the desired sequence of projectivities.

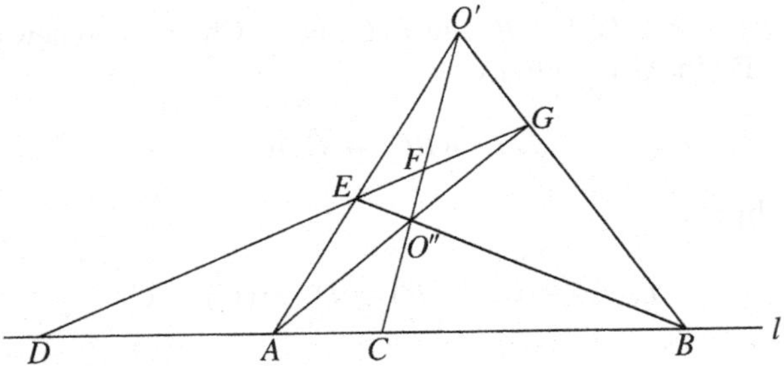


**13** See the solution to Exercise 11.

**15** When constructing $E$ and $F$ we will find that $E$ is between $A$ and $C$ and $F$ is between $B$ and $C$.

## 14.4 HOMOGENEOUS COORDINATES

**1** Suppose that there is a point $P$ that is on every line. Let $l$ and $m$ be two distinct lines (Postulate 14.1.1.4). Let $A$ be on $l$ and $B$ on $m$. Assume that these points are distinct and different from $P$ (Postulate 14.1.1.3). Join $\overline{AB}$. This line is not $l$ and not $m$, but it does contain $P$. Therefore, $A$ and $P$ are on both $\overline{AB}$ and $l$, contradicting (Postulate 14.1.1.1).

**3(b)** The coordinates for $E_2$ are (0, 1, 0), and the coordinates for $E_3$ are (0, 0, 1). We will find the line $ax + by + cz = 0$ that contains both of these points. This means that we must solve the system given by the matrix

$$\begin{bmatrix} 0 & 1 & 0 \\ 0 & 0 & 1 \end{bmatrix}$$

The solution set to this system (in homogeneous coordinates) is

$$\{(s, 0, 0) : s \in \mathbb{R} \text{ and } s \neq 0\}.$$

Therefore, $\overline{E_2E_3} = [1, 0, 0]$.

**5** The equation for the point of intersection is $17a + 23b - 5c = 0$.

**7(a)** In homogeneous coordinates,

$$(0, 0, 5) = (0, 0, 7),$$
$$(1, 1, 1) = (7, 7, 7),$$
$$(7, 7, 0) = (-7, -7, 0).$$

So we have

$$(0, 0, 7) = (7, 7, 7) + (-7, -7, 0).$$

**9(a)** Let $a$, $b$, and $c$ be nonzero, real numbers.

- $a \sim a$ because $a^2 > 0$.
- Let $a \sim b$. This means that $ab > 0$. This implies that $ba > 0$, so $b \sim a$.
- Assume that $a \sim b$ and $b \sim c$. This means that $ab > 0$ and $bc > 0$. We have two possibilities: First, $a$, $b$, and $c$ are all positive. This implies that $ac > 0$. Second, $a$, $b$, and $c$ are all negative, which implies that $ac > 0$. Hence, in either case we have $a \sim c$.

**11(a)** Let $(a, b), (c, d), (e, f) \in \mathbb{Z}\backslash\{0\} \times \mathbb{Z}\backslash\{0\}$.

- Since $ab = ab$, $(a, b) \sim (a, b)$.
- Suppose that $(a, b) \sim (c, d)$. This means that $ad = bc$. From this we conclude that $(c, d) \sim (a, b)$ because $cb = da$.
- Let $(a, b) \sim (c, d)$ and $(c, d) \sim (e, f)$. Then, $ad = bc$ and $cf = de$. This implies that

  $$adcf = bcde,$$

  and since $d \neq 0$ and $c \neq 0$,

  $$af = be.$$

  Hence, $(a, b) \sim (e, f)$.

**11(c)** This is the set of equivalence classes of $(\mathbb{Z}\backslash\{0\} \times \mathbb{Z}\backslash\{0\})$ under $\sim$. Each of these classes is of the form

$$[(a, b)] = \{(x, y) : xb = ay\}.$$

Observe that two ordered pairs are equivalent in the same manner that two fractions are equivalent, namely, that $a/b = c/d$ implies that $ad = bc$. Thus, $(\mathbb{Z}\backslash\{0\} \times \mathbb{Z}\backslash\{0\})/\sim$ can be viewed as the set of nonzero fractions.

**13** No, because the ordered pair $(1, 1)$ is not an element of

$$\bigcup\{[n, n+1] \times (n, n+1) : n \in \mathbb{Z}\}.$$

**15(a)** The matrix for $\psi$ is

$$\begin{bmatrix} 4 & 10 & -14 \\ -4 & 25 & -14 \\ 16 & 5 & -28 \end{bmatrix}$$

**17** Since $\overline{EE_3} = [1, -1, 0]$ and $\overline{E_1E_2} = [0, 0, 1]$, the homogeneous coordinates for the point $O$ are $(-1, 1, 0)$. Since $A$ is on $\overline{E_2E_3} = [1, 0, 0]$, we have that

$$A = (0, a_1, a_2),$$

where $a_1$ and $a_2$ are not both zero. Also, since $B$ is on $\overline{E_1E_3} = [0, 1, 0]$, we have that

$$B = (b_1, 0, b_2),$$

where $b_1$ and $b_2$ are not both zero. In order for $A$, $B$, and $O$ to be collinear, by Corollary 14.4.3,

$$\begin{vmatrix} -1 & 1 & 0 \\ 0 & a_1 & a_2 \\ b_1 & 0 & b_2 \end{vmatrix} = 0,$$

but this is equivalent to the equation

$$a_2b_1 - a_1b_2 = 0.$$

# CHAPTER 15

# FELIX KLEIN

## 15.1 GROUP THEORY

**1(a)** It is not a binary operation. It is not closed.

**1(c)** It is not a binary operation. It is not closed.

**1(e)** It is a binary operation.

**3(a)** The operation is closed because if $\mathscr{A}$ and $\mathscr{B}$ are subsets of $\mathscr{S}$, then $\mathscr{A} \cap \mathscr{B}$ is a subset of $\mathscr{S}$. To see that $*$ is a function, let $\mathscr{A}$, $\mathscr{A}'$, $\mathscr{B}$, $\mathscr{B}' \in P(\mathscr{S})$. Suppose that $\mathscr{A} = \mathscr{A}'$ and $\mathscr{B} = \mathscr{B}'$. Then, clearly, $\mathscr{A} \cap \mathscr{B} = \mathscr{A}' \cap \mathscr{B}'$.

**5(b)** Define $*$ by $x * y = x + y + 2$. We note that $*$ is a binary operation. As for the group properties:

- Let $a,\ b,\ c \in \mathbb{Z}$. We calculate:

$$a * (b * c) = a * (b + c + 2) = a + (b + c + 2) + 2$$

and

$$(a * b) * c = (a * b) + c + 2 = (a + b + 2) + c + 2.$$

*Revolutions of Geometry.* By Michael O'Leary

Since $a + (b + c + 2) + 2 = (a + b + 2) + c + 2$, $*$ is associative.

- $-2$ is the identity of the group because
$$a * (-2) = a + (-2) + 2 = a = (-2) + a + 2 = (-2) * a.$$
- Let $a \in \mathbb{Z}$. Then $-a - 4$ is the inverse of $a$ in the group because
$$a * (-a - 4) = a + (-a - 4) + 2 = -2$$

and
$$(-a - 4) * a = (-a - 4) + a + 2 = -2.$$

Therefore, $(\mathbb{Z}, *)$ is a group.

**7(a)** It is well known that matrix addition is a commutative binary operation, so we will only check the group properties. Let

$$\begin{bmatrix} a_{1,1} & \cdots & a_{1,n} \\ \vdots & & \vdots \\ a_{n,1} & \cdots & a_{n,n} \end{bmatrix}, \begin{bmatrix} b_{1,1} & \cdots & b_{1,n} \\ \vdots & & \vdots \\ b_{n,1} & \cdots & b_{n,n} \end{bmatrix}, \begin{bmatrix} c_{1,1} & \cdots & c_{1,n} \\ \vdots & & \vdots \\ c_{n,1} & \cdots & c_{n,n} \end{bmatrix} \in M_n.$$

We first check that matrix addition is associative. The sum

$$\begin{bmatrix} a_{1,1} & \cdots & a_{1,n} \\ \vdots & & \vdots \\ a_{n,1} & \cdots & a_{n,n} \end{bmatrix} + \left( \begin{bmatrix} b_{1,1} & \cdots & b_{1,n} \\ \vdots & & \vdots \\ b_{n,1} & \cdots & b_{n,n} \end{bmatrix} + \begin{bmatrix} c_{1,1} & \cdots & c_{1,n} \\ \vdots & & \vdots \\ c_{n,1} & \cdots & c_{n,n} \end{bmatrix} \right)$$

equals

$$\begin{bmatrix} a_{1,1} & \cdots & a_{1,n} \\ \vdots & & \vdots \\ a_{n,1} & \cdots & a_{n,n} \end{bmatrix} + \begin{bmatrix} b_{1,1} + c_{1,1} & \cdots & b_{1,n} + c_{1,n} \\ \vdots & & \vdots \\ b_{n,1} + c_{n,1} & \cdots & b_{n,n} + c_{n,n}, \end{bmatrix},$$

and this in turn equals

$$\begin{bmatrix} a_{1,1} + (b_{1,1} + c_{1,1}) & \cdots & a_{1,n} + (b_{1,n} + c_{1,n}) \\ \vdots & & \vdots \\ a_{n,1} + (b_{n,1} + c_{n,1}) & \cdots & a_{n,n} + (b_{n,n} + c_{n,n}) \end{bmatrix}.$$

This last matrix equals

$$\begin{bmatrix} (a_{1,1} + b_{1,1}) + c_{1,1} & \cdots & (a_{1,n} + b_{1,n}) + c_{1,n} \\ \vdots & & \vdots \\ (a_{n,1} + b_{n,1}) + c_{n,1} & \cdots & (a_{n,n} + b_{n,n}) + c_{n,n} \end{bmatrix}$$

because addition of real numbers is associative. Therefore, we may back out of the calculation and arrive at

$$\left( \begin{bmatrix} a_{1,1} & \cdots & a_{1,n} \\ \vdots & & \vdots \\ a_{n,1} & \cdots & a_{n,n} \end{bmatrix} + \begin{bmatrix} b_{1,1} & \cdots & b_{1,n} \\ \vdots & & \vdots \\ b_{n,1} & \cdots & b_{n,n} \end{bmatrix} \right) + \begin{bmatrix} c_{1,1} & \cdots & c_{1,n} \\ \vdots & & \vdots \\ c_{n,1} & \cdots & c_{n,n} \end{bmatrix}.$$

The zero matrix is the identity since

$$\begin{bmatrix} a_{1,1} & \cdots & a_{1,n} \\ \vdots & & \vdots \\ a_{n,1} & \cdots & a_{n,n} \end{bmatrix} + \begin{bmatrix} 0 & \cdots & 0 \\ \vdots & & \vdots \\ 0 & \cdots & 0 \end{bmatrix} = \begin{bmatrix} a_{1,1} & \cdots & a_{1,n} \\ \vdots & & \vdots \\ a_{n,1} & \cdots & a_{n,n} \end{bmatrix},$$

and the inverse of

$$\begin{bmatrix} a_{1,1} & \cdots & a_{1,n} \\ \vdots & & \vdots \\ a_{n,1} & \cdots & a_{n,n} \end{bmatrix}$$

is

$$\begin{bmatrix} -a_{1,1} & \cdots & -a_{1,n} \\ \vdots & & \vdots \\ -a_{n,1} & \cdots & -a_{n,n} \end{bmatrix}$$

because

$$\begin{bmatrix} a_{1,1} & \cdots & a_{1,n} \\ \vdots & & \vdots \\ a_{n,1} & \cdots & a_{n,n} \end{bmatrix} + \begin{bmatrix} -a_{1,1} & \cdots & -a_{1,n} \\ \vdots & & \vdots \\ -a_{n,1} & \cdots & -a_{n,n} \end{bmatrix} = \begin{bmatrix} 0 & \cdots & 0 \\ \vdots & & \vdots \\ 0 & \cdots & 0 \end{bmatrix}.$$

**7(c)** $M_n$ is not a group under matrix multiplication because not every matrix has a multiplicative inverse.

**9** We first prove that $\phi$ is one-to-one. Let $x,\ y \in \mathcal{G}$ and suppose that $\phi(x) = \phi(y)$. This means that $x^{-1} = y^{-1}$. Therefore, $(x^{-1})^{-1} = (y^{-1})^{-1}$, from which it follows that $x = y$ (Theorem 15.1.3). To see that $\phi$ is onto, let $y \in \mathcal{G}$. Then, $y^{-1} \in \mathcal{G}$ and

$$\phi(y^{-1}) = (y^{-1})^{-1} = y.$$

**11** Let $a,\ b \in \mathcal{G}$. Since $(a * b)^2 = a^2 * b^2$,

$$\begin{aligned} (a * b) * (a * b) &= (a * a) * (b * b) \\ a * [b * (a * b)] &= a * [a * (b * b)] \\ b * (a * b) &= a * (b * b) \\ (b * a) * b &= (a * b) * b \\ b * a &= a * b. \end{aligned}$$

The third equation follows from the second by operating both sides of the equation on the left by $a^{-1}$, and the fifth equation follows from the fourth by operating both sides on the right by $b^{-1}$.

**12(a)** $\{e\}, \{e, a\}, \{e, b\}, \{e, c\}, \{e, a, b, c\}$

**12(d)** $\{[0]_2, [0]_3\}, \mathbb{Z}_2 \times \{[0]_3\}, \{[0]_2\} \times \mathbb{Z}_3, \mathbb{Z}_2 \times \mathbb{Z}_3.$

**13** Let $\mathcal{H} = \{f \in \mathcal{F} : f(0) = 0\}$. Take $f, g \in \mathcal{H}$. Then,

$$(f + g)(0) = f(0) + g(0) = 0,$$

so $f = g \in \mathcal{H}$. Also, the identity map, $f(x) = 0$ for all $x$, is an element of $\mathcal{H}$. Lastly, if $f \in \mathcal{H}$, then $-f \in \mathcal{H}$ because $(-f)(0) = -f(0) = 0$.

**15(a)** Both $2\mathbb{Z} = \{2n : n \in \mathbb{Z}\}$ and $3\mathbb{Z} = \{3n : n \in \mathbb{Z}\}$ are groups under addition. However, addition is not a binary operation on $2\mathbb{Z} \cup 3\mathbb{Z}$ because $2 + 3$ is not an element of the union.

**17** To ease our notation, let $xy = x * y$.

- First, take $c, d \in \mathcal{H}$ and calculate for any $b \in \mathcal{S}$,

$$(cd)b = c(db) = c(bd) = (cb)d = (bc)d = b(cd).$$

- Second, $e \in \mathcal{H}$ because $eb = be$ for all $b \in \mathcal{S}$.

- Finally, let $c \in \mathcal{H}$, so for all $b \in \mathcal{S}$,

$$bc^{-1} = (c^{-1}c)(bc^{-1}) = c^{-1}[(bc)c^{-1})] = c^{-1}b.$$

  Thus, $c^{-1} \in \mathcal{H}$.

Therefore, $\mathcal{H}$ is a subgroup of $\mathcal{G}$.

**19** Let $\mathcal{H}$ be a subgroup of the group $\mathcal{G}$ and $\mathcal{H}'$ be a subgroup of the group $\mathcal{G}'$. Clearly, $\mathcal{H} \times \mathcal{H}' \subseteq \mathcal{G} \times \mathcal{G}'$, so we check the three conditions:

- Let $(a, a'), (b, b') \in \mathcal{H} \times \mathcal{H}'$. Since both $\mathcal{H}$ and $\mathcal{H}'$ are subgroups, $a * b \in \mathcal{H}$ and $a' * b' \in \mathcal{H}'$. Hence, $(a, a') \cdot (b, b') \in \mathcal{H} \times \mathcal{H}'$.

- Denote the identity of $\mathcal{G}$ by $e$ and the identity of $\mathcal{G}'$ by $e'$. Then, $(e, e')$ is the identity of $\mathcal{G} \times \mathcal{G}'$. Furthermore, $(e, e') \in \mathcal{H} \times \mathcal{H}'$ because $e \in \mathcal{H}$ and $e' \in \mathcal{H}'$.

- Lastly, let $(a, b) \in \mathcal{H} \times \mathcal{H}'$. Since $a \in \mathcal{H}$ and $b \in \mathcal{H}'$, $a^{-1} \in \mathcal{H}$ and $b^{-1} \in \mathcal{H}'$. Hence, $(a, b)^{-1} = (a^{-1}, b^{-1}) \in \mathcal{H} \times \mathcal{H}'$.

Therefore, $\mathcal{H} \times \mathcal{H}'$ is a subgroup of $\mathcal{G} \times \mathcal{G}'$.

**21** Let $\mathcal{G}$ be a group and $a \in \mathcal{G}$. Let $k$ be a positive integer. Write $\mathcal{H} = \{a^{nk} : n \in \mathbb{Z}\}$. Certainly, $\mathcal{H} \subseteq \langle a \rangle$.

- Let $n, m \in \mathbb{Z}$ and note that $a^{nk}a^{mk} = a^{k(n+m)}$ and $n + m$ is an integer. Hence, $\mathcal{H}$ is closed.

- Since $e = a^{0 \cdot k}$, $e \in \mathcal{H}$.

- Let $n \in \mathbb{Z}$. Then, $a^{-nk}$ is the inverse of $a^{nk}$, and $a^{-nk} \in \mathcal{H}$.

Hence, $\mathcal{H}$ is a subgroup of $\langle a \rangle$.

**22** Let $\mathcal{G}$ be a simple group. If $\mathcal{G} = \{e\}$, then the group is cyclic with generator $e$. If $\mathcal{G} \neq \{e\}$, take $a \in \mathcal{G}$ such that $a \neq e$. We know that $\langle a \rangle$ is a group and thus a subgroup of $\mathcal{G}$. Since $\mathcal{G}$ can have no proper subgroups, $\langle a \rangle$ must equal $\mathcal{G}$. Therefore, $\mathcal{G}$ is cyclic with generator $a$.

## 15.2 TRANSFORMATION GROUPS

**1** Let $(\mathcal{P}, \mathcal{L})$ be a projective plane. We will check that $(\mathbb{A}(\mathcal{P}), \mathbb{L}(\mathcal{L}))$ satisfies Postulate 15.2.1.

- Postulate 15.2.1 follows because of Postulate 14.1.1.1.
- We now show that every line in $\mathbb{L}(\mathcal{L})$ has at least two points. Let $l \in \mathbb{L}(\mathcal{L})$ have equation

$$ax + by + c = 0,$$

where $[a, b, c] \in \mathcal{L}$ and this is not $[0, 0, 1]$. By Postulate 14.1.1.3, there are at least three points on $[a, b, c]$. At most one is on $[0, 0, 1]$, so there are at least two points that have homogenous coordinates $(u_1, v_1, 1)$ and $(u_2, v_2, 1)$. Therefore, $u_1 \neq u_2$ or $v_1 \neq v_2$, and both

$$au_1 + bv_1 + c = 0$$

and

$$au_2 + bv_2 + c = 0.$$

We have thus found two distinct points on $l$, namely $(u_1, v_1)$ and $(u_2, v_2)$.

- We now need to show that there are at least three noncollinear points. We know that we have points in $\mathcal{P}$ that have homogeneous coordinates $(0, 0, 1)$, $(0, 1, 1)$ and $(1, 1, 1)$. These correspond to points in $\mathbb{A}(\mathcal{P})$, namely $(0, 0)$, $(0, 1)$, and $(1, 1)$. These three points are not collinear.
- Let $l$ be a line in $\mathbb{L}(\mathcal{L})$ and $P$ a point in $\mathbb{A}(\mathcal{P})$ that is not on $l$. If both $m$ and $n$ are on $P$ and parallel to $l$, then $m$ and $n$ both meet at the same ideal point in $\mathcal{P}$, contradicting Postulate 14.1.1.1. Therefore, there is at most one parallel to $l$ through $P$. There is at least one parallel because $l$ must intersect $[0, 0, 1]$ at some $\Omega$ (Postulate 14.1.1.2). Let $ax + by + cz = 0$ be the equation for $\overline{P\Omega}$. This means that $ax + by + c = 0$ is the equation for the line through $P$ parallel to $l$. Hence, the fourth postulate is satisfied.

**3(a)** By the definition of $(\mathcal{P}_A, \mathcal{L}_A)$, every two points are on exactly one line, and every line contains at least two points. This confirms the first two postulates. The third postulate is seen to be true since $A$, $B$, and $C$ are not collinear. Lastly, given any line, there are exactly two points not on the line. The line through these two points will be the only line parallel to the given line. For example, the points $C$ and $D$ are the only points not on line $\{A, B\}$, and $\{C, D\}$ is the only line not incident with $\{A, B\}$.

**5(a)** $B = (0, 1, 1)$, $D = (1, 1, 0)$, $F = (1, 0, 1)$.

**5(c)** $B = (0, 1)$, $C = (1, 1)$, $E = (0, 0)$, and $F = (1, 0)$. $A$, $D$, and $G$ do not have nonhomogeneous coordinates.

**5(f)** The affinity is

$$\phi(x, y) = \begin{bmatrix} 0 & 1 \\ 1 & 0 \end{bmatrix} \begin{bmatrix} x \\ y \end{bmatrix}.$$

**5(g)** The desired collineation is

$$\phi : \mathbb{A} \to \mathcal{P}_A,$$

which is defined by

$$\phi(0, 0) = E, \phi(1, 0) = F, \phi(0, 1) = B, \text{ and } \phi(1, 1) = C.$$

**7** No for 10 points. Yes for 16.

**9** The affine plane of Exercise 3 shows consistency. The two models in the diagram each satisfy exactly three of the affine postulates. The first does not satisfy the first postulate, and the second does not satisfy the fourth affine postulate. The empty set does not satisfy the third postulate but vacuously satisfies the others. However, Postulates 15.2.1 are not independent because the second postulate can be proven from the other three.

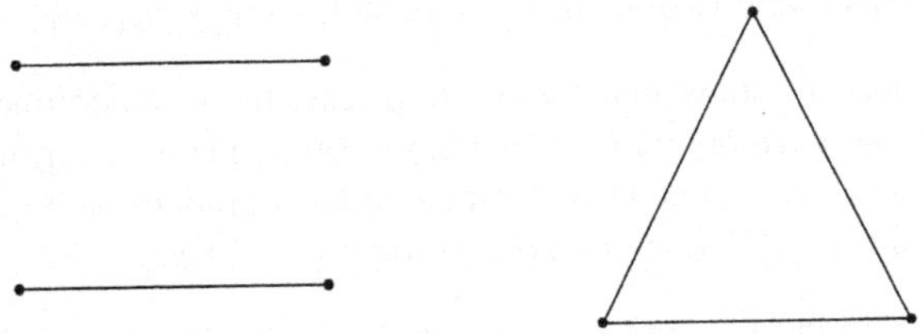

**11** The homogeneous coordinates for (4, 10) are of the form $(4s, 10s, s)$ with $s \neq 0$, and the nonhomogeneous coordinates for (4, 10, 7) are (4/7, 10/7).

**13** We can use this equation:

$$\phi(x, y) = \begin{bmatrix} -6 & 10 \\ 6 & 0 \end{bmatrix} \begin{bmatrix} x \\ y \end{bmatrix} + \begin{bmatrix} 0 \\ -6 \end{bmatrix}.$$

**14(c)** With our definition of parallel, the set of dilatations cannot be a group because the identity map is not a dilatation.

**15** Let $\mathcal{A}$ be the set of affine translations. *Note:* None of our dilatations have a fixed point because of our definition of *parallel*. Sometimes in the study of affine geometry the phrase *l is parallel to m* means that $l$ and $m$ do not intersect or $l$ and $m$ are the same line. This is not the case here.

- Let $\phi$, $\psi \in \mathcal{A}$. If either function is the identity, the composition is an affine transformation, so assume the neither $\phi$ nor $\psi$ is the identity map. Let $l$ be a line in the affine plane. Then, $\psi[l]$ is parallel to $l$, and $\phi[\psi[l]]$ is parallel to

$\psi[l]$. Hence, $\phi[\psi[l]]$ is either parallel or equal to $l$, showing that $\phi \circ \psi$ is an affine translation.

- The identity is an affine translation by definition.
- Let $\phi \in \mathcal{A}$. If $\phi$ is the identity map, then $\phi^{-1} = \phi$, so assume that $\phi$ is not the identity. Since $\phi$ is a transformation, $\phi^{-1}$ is a function. Let $l$ be a line. Then, $\phi^{-1}[l] = m$ for some line $m$. Hence, $\phi[m] = l$, so $m$ is parallel to $l$. This implies that $\phi^{-1} \in \mathcal{A}$.

**17(b)** The set of shears does not form a group under composition because the identity is not a shear.

## 15.3 THE PRINCIPAL GROUP

**1(a)** We must show that $f$ is one-to-one and onto.

- Let $x, y \in \mathbb{R}$. Assume that $f(x) = f(y)$. This means that
$$x^3 + 1 = y^3 + 1.$$
Hence, $x^3 = y^3$, which implies that $x = y$. So, $f$ is one-to-one.
- Let $y \in \mathbb{R}$. Then,
$$f(\sqrt[3]{x-1}) = (\sqrt[3]{x-1})^3 + 1 = (x-1) + 1 = x.$$
Therefore, $f$ is onto.

**1(b)** For $f : \mathbb{R} \to \mathbb{R}$ to be an isometry, it must be the case that for all $x, y \in \mathbb{R}$,
$$|x - y| = |f(x) - f(y)|.$$
The function defined in this exercise is not an isometry because $|2 - 1| = 1$ but $|f(2) - f(1)| = 7$.

**3** Let $(x, y), (x', y') \in \mathbb{R}^2$. Then,
$$\begin{aligned}
d[f(x, y), f(x', y')] &= d[(x - 7, y + 5), (x' - 7, y' + 5)] \\
&= \sqrt{[(x-7) - (x'-7)]^2 + [(y+5) - (y'+5)]^2} \\
&= \sqrt{(x - x')^2 + (y - y')^2} \\
&= d[(x, y), (x', y')].
\end{aligned}$$

**5** Isometries do preserve area. For example, let $f: \mathbb{R}^2 \to \mathbb{R}^2$ be an isometry. Let $ABCD$ be a rectangle. We know that since $f$ is an isometry,
$$\begin{aligned}
AB &= \|f[\overline{AB}]\|, \\
BC &= \|f[\overline{BC}]\|, \\
CD &= \|f[\overline{CD}]\|, \\
AD &= \|f[\overline{AD}]\|.
\end{aligned}$$

Since isometries are conformal (Theorem 15.3.4), the internal angles of quadrilateral $f(A)f(B)f(C)f(D)$ are right, so the image of $ABCD$ is a rectangle. Therefore,

$$\text{area}(ABCD) = AB \cdot BC = \|f[\overline{AB}]\| \cdot \|f[\overline{BC}]\| = \text{area}(f(A)f(B)f(C)f(D)).$$

**7** Let $f, g : \mathbb{R}^2 \to \mathbb{R}^2$ be two isometries. Let $(x, y), (x', y') \in \mathbb{R}^2$. We calculate:

$$\begin{aligned} d[(f \circ g)(x, y), (f \circ g)(x', y')] &= d[f(g(x, y)), f(g(x', y'))] \\ &= d[g(x, y), g(x', y')] \\ &= d[(x, y), (x', y')]. \end{aligned}$$

The second inequality follows because $f$ is an isometry, and the third follows because $g$ is an isometry.

**9** Yes, because an isometry is determined by three noncollinear points (Theorem 15.3.5).

**11** Let $\sigma$ be a similarity with ratio $r > 0$. Let $U$ be $(1, 0)$ and $V$ be $(0, 1)$. Let $\sigma(U) = U'$ and $\sigma(V) = V'$. Since $U$ and $V$ are one unit from the origin $O$, both $U'$ and $V'$ must lie on the circle with center $O$ and radius $r$. Moreover, since $UV = \sqrt{2}$, we know that $U'V' = r\sqrt{2}$. We first assume that $U'$ is $(r, 0)$. In this case, $V'$ is either $(0, r)$ or $(0, -r)$ because both of these points are on $\odot O$ and are a distance of $r\sqrt{2}$ from $(r, 0)$. Therefore, the matrix for $\sigma$ is

$$\begin{bmatrix} r & 0 \\ 0 & r \end{bmatrix}$$

or

$$\begin{bmatrix} r & 0 \\ 0 & -r \end{bmatrix}.$$

Now, suppose that $\sigma$ is defined so that $U'$ is any point on $\odot O$. Let $\theta$ be the angle that $\overline{OU'}$ makes with the $x$-axis. Then the matrix for $\sigma$ is given by the composition of the rotation matrix with one of the matrices above giving either

$$\begin{bmatrix} r\cos\theta & -r\sin\theta \\ r\sin\theta & r\cos\theta \end{bmatrix}$$

or

$$\begin{bmatrix} r\cos\theta & r\sin\theta \\ r\sin\theta & -r\cos\theta \end{bmatrix}$$

as the matrix for $\sigma$. Composing this similarity with a translation described by $\langle a, b\rangle$, we find the equation for an arbitrary similarity of the plane as either

$$\sigma(x, y) = r\begin{bmatrix} \cos\theta & -\sin\theta \\ \sin\theta & \cos\theta \end{bmatrix}\begin{bmatrix} x \\ y \end{bmatrix} + \begin{bmatrix} a \\ b \end{bmatrix}$$

or

$$\sigma(x, y) = r\begin{bmatrix} \cos\theta & \sin\theta \\ \sin\theta & -\cos\theta \end{bmatrix}\begin{bmatrix} x \\ y \end{bmatrix} + \begin{bmatrix} a \\ b \end{bmatrix}.$$

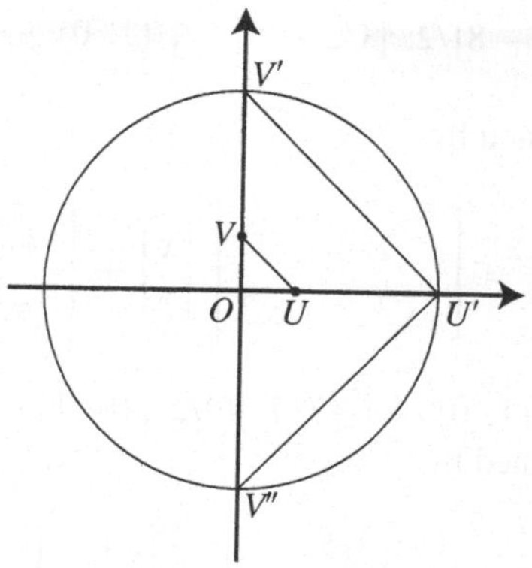


**12(b)** Let $\eta$ be a homothety with center $C$ and radius $r$ and let $\zeta$ be a homothety with center $C$ and radius $s$. Then,

$$d[(\eta \circ \zeta)(A),\ C] = d[\eta(\zeta(A)),\ C] = rd[\zeta(A),\ C] = rsd[A,\ C].$$

Therefore, $\eta \circ \zeta$ is a homothety with center $C$ and radius $rs$, the product of the radii of the original homotheties.

**13** Let $\sigma : \mathbb{R}^2 \to \mathbb{R}^2$ be a similarity with ratio $r > 0$. Since $\sigma$ is a transformation, $\sigma^{-1}$ is also a transformation. Let $A' = \sigma(A)$ and $B' = \sigma(B)$ for any $A,\ B \in \mathbb{R}^2$. Then $A = \sigma^{-1}(A')$ and $B = \sigma^{-1}(B')$. Furthermore, since $A'B' = rAB$, we have that $AB = \frac{1}{r}A'B'$. Therefore, $\sigma^{-1}$ is a similarity with ratio $1/r$.

**15** The set of dilations is not a subgroup of the group of similarities because the inverse of a dilation with radius $r > 1$ is a contraction with radius $1/r$. The set of contractions is also not a subgroup because it does not contain the identity map.

**17** True. For example, let $AB = 3$ be the radius of a given circle. Its area is $9\pi$. Let $\sigma$ be a similarity of radius $r$. The image of the radius of the given circle under $\sigma$ has length $3r$, and its area is $9r^2\pi$.

## 15.4 ISOMETRIES OF THE PLANE

**1(a)** $(0,\ 0) \mapsto (0,\ -8),\ (1,\ 0) \mapsto (1,\ -8),\ (0,\ 1) \mapsto (0,\ -9)$
This function can be defined by

$$\mu(x,\ y) = \begin{bmatrix} 1 & 0 \\ 0 & -1 \end{bmatrix} \begin{bmatrix} x \\ y \end{bmatrix} + \begin{bmatrix} 0 \\ -8 \end{bmatrix}.$$

**1(c)** $(0,\ 0) \mapsto (2,\ -3),\ (1,\ 0) \mapsto (3,\ -3),\ (0,\ 1) \mapsto (2,\ -2)$
This function can be defined by

$$\tau(x,\ y) = \begin{bmatrix} 1 & 0 \\ 0 & 1 \end{bmatrix} \begin{bmatrix} x \\ y \end{bmatrix} + \begin{bmatrix} 2 \\ -3 \end{bmatrix}.$$

**1(f)** $(0, 0) \mapsto ([-7\sqrt{2} - 8]/2, [\sqrt{2} - 6]/2)$, $(1, 0) \mapsto (-4\sqrt{2} - 4, \sqrt{2} - 3)$, $(0, 1) \mapsto (-4\sqrt{2} - 4, -3)$
This function can be defined by

$$\rho(x, y) = \begin{bmatrix} -\frac{\sqrt{2}}{2} & -\frac{\sqrt{2}}{2} \\ \frac{\sqrt{2}}{2} & -\frac{\sqrt{2}}{2} \end{bmatrix} \begin{bmatrix} x \\ y \end{bmatrix} + \begin{bmatrix} \frac{-7\sqrt{2}-8}{2} \\ \frac{\sqrt{2}-6}{2} \end{bmatrix}.$$

**1(h)** $(0, 0) \mapsto (-2, 9)$, $(1, 0) \mapsto (-7/5, 49/5)$, $(0, 1) \mapsto (-6/5, 42/5)$
This function can be defined by

$$\gamma(x, y) = \begin{bmatrix} 3/5 & 4/5 \\ 4/5 & -3/5 \end{bmatrix} \begin{bmatrix} x \\ y \end{bmatrix} + \begin{bmatrix} -2 \\ 9 \end{bmatrix}.$$

**2(b)** See above.

**3(b)** Reflection through a diagonal, reflection through the midpoints of opposite sides, rotations of measure $k\pi/2$ $(k \in \mathbb{Z})$.

**3(d)** Rotations of measure $k\pi$ $(k \in \mathbb{Z})$. There are more if the parallelogram is a rectangle.

**5** We prove sufficiency. Let $l$ be the mirror with angle of inclination $\theta$ and passing through the point $(a, b)$. A reflection through $l$ is equivalent to the following sequence of transformations:

- Translate using $\langle -a, -b \rangle$ so that $l$ is a line through the origin.
- Rotate the plane through an angle of $-\theta$ with the origin as the center.
- Reflect through the $x$-axis.
- Rotate the plane through an angle of $\theta$ with the origin as the center.
- Translate back using $\langle a, b \rangle$.

Let $P(x, y)$ be a point in $\mathbb{R}^2$. The image of this point under the translation described by $\langle a, b \rangle$ is $(x - a, y - b)$. The composition of a rotation about the origin with angle $\theta$, a reflection through the $x$-axis, and a rotation about the origin with angle $-\theta$ is given by

$$\begin{bmatrix} \cos\theta & -\sin\theta \\ \sin\theta & \cos\theta \end{bmatrix} \begin{bmatrix} 1 & 0 \\ 0 & -1 \end{bmatrix} \begin{bmatrix} \cos(-\theta) & -\sin(-\theta) \\ \sin(-\theta) & \cos(-\theta) \end{bmatrix} = \begin{bmatrix} \cos 2\theta & \sin 2\theta \\ \sin 2\theta & -\cos 2\theta \end{bmatrix}.$$

Therefore, letting $\mu$ be the reflection through $l$, we conclude that

$$\mu(x, y) = \begin{bmatrix} \cos 2\theta & \sin 2\theta \\ \sin 2\theta & -\cos 2\theta \end{bmatrix} \begin{bmatrix} x - a \\ y - b \end{bmatrix} + \begin{bmatrix} a \\ b \end{bmatrix}.$$

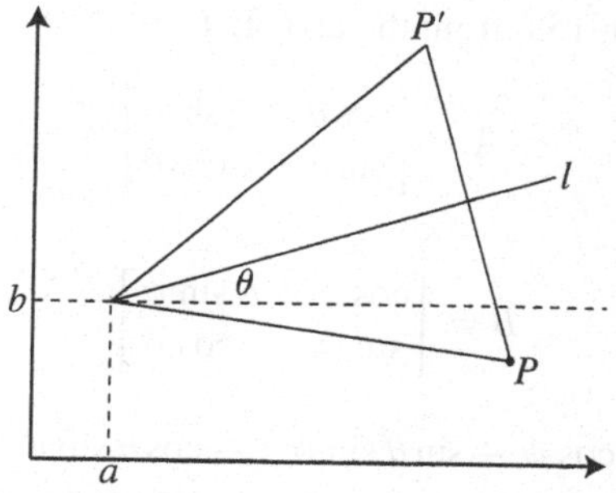


**7** We will prove necessity for Theorems 15.4.7. Let $\tau$ be the composition of reflections $\mu_1$ and $\mu_2$ with parallel mirrors. By Theorem 15.4.3,

$$\mu_1(\mathbf{x}) = A\mathbf{x} + \mathbf{b}$$

and

$$\mu_2(\mathbf{x}) = A\mathbf{x} + \mathbf{c},$$

where $\mathbf{b}$, $\mathbf{c} \in \mathbb{R}^2$ and for some $\theta$,

$$A = \begin{bmatrix} \cos\theta & \sin\theta \\ \sin\theta & -\cos\theta \end{bmatrix}.$$

Notice that $A^2$ is the identity matrix. Hence

$$\tau(\mathbf{x}) = (\mu_1 \circ \mu_2)(\mathbf{x}) = \mu_1(A\mathbf{x} + \mathbf{c}) = A^2\mathbf{x} + A\mathbf{c} + \mathbf{d} = \mathbf{x} + A\mathbf{c} + \mathbf{d}.$$

Hence, $\tau$ is a translation because $A\mathbf{c} + \mathbf{d} \in \mathbb{R}^2$.

**9** Let $\triangle ABC$ be an isosceles triangle with base $\overline{BC}$. Let $D$ be the midpoint of the base. Then the triangle is symmetric about the line $\overline{AD}$, which implies that the base angles must be congruent.

**11** Compose three reflections of the form given in Theorem 15.4.3 and compare the result with that in the theorem and Equation 15.7. Notice that the two forms differ only in the vector that is added to the result of the application of the rotation matrix. Namely, in the reflection,

$$\begin{bmatrix} a \\ b \end{bmatrix}$$

is added, while in the glide reflection,

$$\begin{bmatrix} a + \|\mathbf{v}\| \cos\theta \\ b + \|\mathbf{v}\| \sin\theta \end{bmatrix}$$

is added.

**13** Actually, the composition of two homotheties is a homothety. However, a translation is a homothety with ratio 1.

**15** We will use Equation 15.6 from the text. Let

$$A = \begin{bmatrix} \cos\theta & -\sin\theta \\ \sin\theta & \cos\theta \end{bmatrix}$$

and

$$B = \begin{bmatrix} \cos\psi & -\sin\psi \\ \sin\psi & \cos\psi \end{bmatrix},$$

so

$$AB = \begin{bmatrix} \cos\theta\cos\psi - \sin\theta\sin\psi & -\cos\theta\sin\psi - \sin\theta\cos\psi \\ \sin\theta\cos\psi + \cos\theta\sin\psi & -\sin\theta\sin\psi + \cos\theta\cos\psi \end{bmatrix}$$

Therefore,

$$AB = \begin{bmatrix} \cos(\theta+\psi) & -\sin(\theta+\psi) \\ \sin(\theta+\psi) & \cos(\theta+\psi) \end{bmatrix}.$$

Now let $\rho_1$ and $\rho_2$ be rotations about $\mathbf{a} = (a, b)$ such that

$$\rho_1(\mathbf{x}) = A(\mathbf{x} - \mathbf{a}) + \mathbf{a}$$

and

$$\rho_2(\mathbf{x}) = B(\mathbf{x} - \mathbf{a}) + \mathbf{a}.$$

Hence,

$$(\rho_1 \circ \rho_2)(\mathbf{x}) = \rho_1(B(\mathbf{x} - \mathbf{a}) + \mathbf{a}) = A[B(\mathbf{x} - \mathbf{a}) + \mathbf{a} - \mathbf{a}] + \mathbf{a} = AB(\mathbf{x} - \mathbf{a}) + \mathbf{a}.$$

This means that $\rho_1 \circ \rho_2$ is the rotation about $(a, b)$ with angle $\theta + \psi$.

**17** No, because the identity is not a glide reflection.

**19(a)** The equation of the image is

$$y = -\frac{29}{2}x + \frac{495}{22}.$$

**19(c)** The equation of the image is

$$y = \frac{1 + 2\sqrt{3}}{-2 + \sqrt{3}}x + \frac{-5}{-2 + \sqrt{3}}.$$

**21(a)** The lines of symmetry are as illustrated:

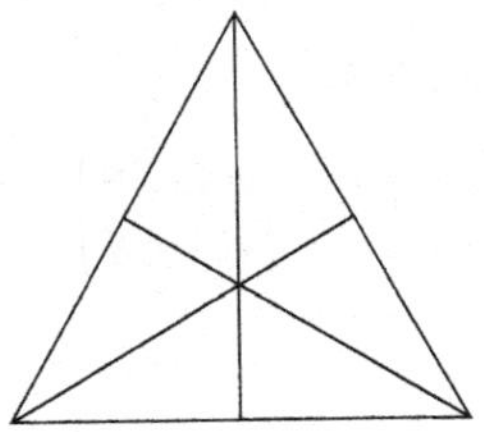

**21(c)** A parallelogram with no right angles has no lines of symmetry.

## 15.5 CONSISTENCY OF EUCLIDEAN GEOMETRY

**1(b)** The Euclidean model shows that the order axioms are consistent, so we need to find four models such that in each one exactly three of the order axioms are satisfied.

- In the first model, the numbers of

$$l = \left\{ \frac{1}{2^n} : n = 0, 1, 2, \ldots \right\}$$

  are the *points*, and the set $l$ is interpreted as the only *line* of the model. A point is *on* $l$ if it is an element of $l$, and $A - B - C$ means

$$A > B > C.$$

  Since $A > B > C$ implies that $C > B > A$ is false, the first axiom fails to hold in this model. Axioms 2 and 3 hold, and Pasch's axiom holds vacuously because there are no triangles.

- The lines and points of the model are as in the diagram, and *on* and *between* have their regular meaning. In this model, there is no point beyond $A$ and $B$, so the second axiom is false. The other three axioms hold vacuously.

- *Points* are points on a sphere, *lines* are great circles, *incidence* has its standard meaning, and $A - B - C$ means that $A$, $B$, and $C$ are on a great circle with $B$ between $A$ and $C$. All of the axioms are satisfied in this model except for the third because $A - B - C$ implies that $B - C - A$ and $C - A - B$.

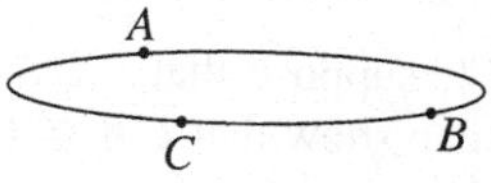


- This last model has only four *lines* as illustrated. *Points* are standard Euclidean points, but the lines intersect only at the indicated points. Specifically, $l$ intersects $\overline{AB}$ only at $D$. *Incidence* and *between* have their standard meaning. With this interpretation, the first three axioms hold in this model, but Pasch's Axiom does not.

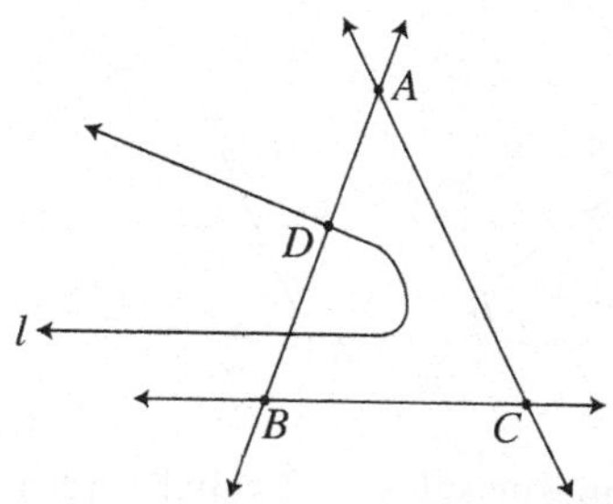


**3** $\overline{AB}$ is parallel to $\overline{CD}$, $\overline{CE}$, and $\overline{DE}$; $\overline{AC}$ is parallel to $\overline{BD}$, $\overline{BE}$, and $\overline{DE}$; $\overline{AD}$ is parallel to $\overline{BC}$, $\overline{BE}$, and $\overline{CE}$; $\overline{AE}$ is parallel to $\overline{BC}$, $\overline{BD}$, and $\overline{CD}$; $\overline{BC}$ is parallel to $\overline{DE}$; $\overline{BD}$ is parallel to $\overline{CE}$; and $\overline{BE}$ is parallel to $\overline{CD}$.

**9** $A$ and $B$ are on opposite sides of $\overline{CD}$ if there exists $E$ on $\overline{CD}$ such that $A - E - B$. The two points are on the same side of $\overline{CD}$ if no such $E$ exists.

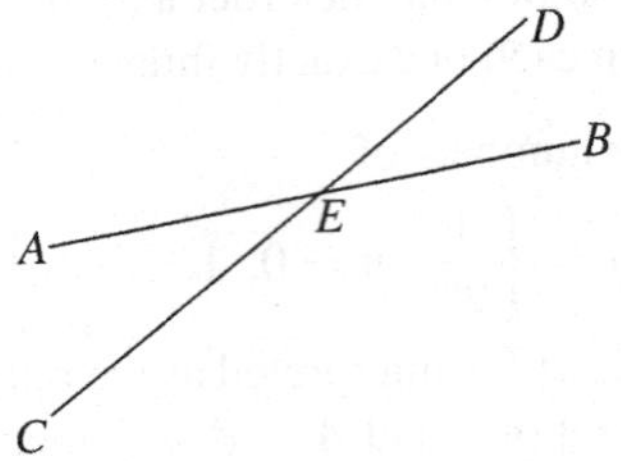


**10(a)** Let $A$ and $B$ be two points. By Axiom 15.5.1.3, there exists point $D$ not on $\overline{AB}$. Join $\overline{AD}$ (Axiom 15.5.1.1) and extend it to a point $E$ such that $A - D - E$ (Axiom 15.5.3.3). Using the same argument, we can join $\overline{EB}$ and extend it to a point $F$ such that $E - B - F$. Therefore, by Axiom 15.5.1.2 and Pasch's Axiom, there exists $C$ on $\overline{AB}$ such that $D - C - B$.

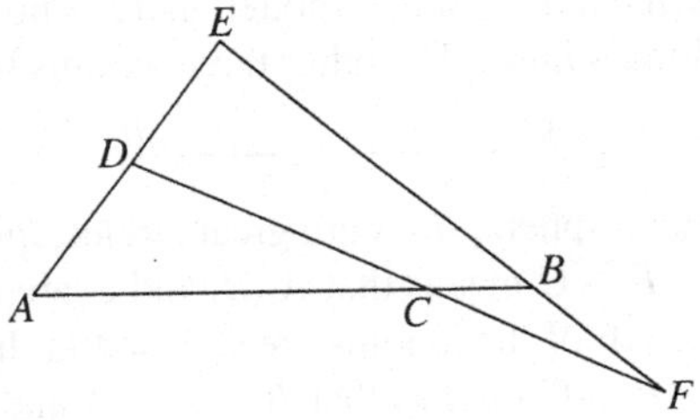


**10(e)** In $\triangle ABC$ and $\triangle A'B'C'$, suppose that $\sphericalangle CAB \cong \sphericalangle C'A'B'$, $\overline{AB} \cong \overline{A'B'}$, and $\sphericalangle ABC \cong A'B'C'$. If we can show that $\overline{CB} \cong \overline{C'B'}$, we can conclude that the triangles are congruent [SAS, Exercise 10(d)], so assume that $\overline{CB}$ is not congruent to $\overline{C'B'}$. Without loss of generality, assume that $CB > C'B'$. We can then find $D$ on $\overline{CB}$ such that $\overline{BD} \cong \overline{B'C'}$ (Axiom 15.5.7.1). Therefore, $\triangle ABD \cong \triangle A'B'C'$ (SAS), and then $\sphericalangle DAB \cong \sphericalangle C'A'B'$. However, this implies that $\sphericalangle DAB \cong \sphericalangle CAB$, in contradiction to Axiom 10(d).4.

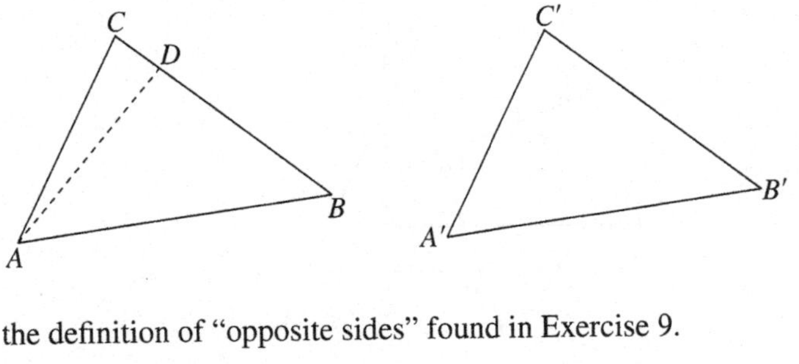


**10(i)** Use the definition of "opposite sides" found in Exercise 9.

Printed and bound by CPI Group (UK) Ltd, Croydon, CR0 4YY
21/09/2026
02220250-0001